T0339919

Authorization and Access Control

Authorization and Access Control
Foundations, Frameworks, and Applications

Parikshit N. Mahalle
Shashikant S. Bhong
Gitanjali R. Shinde

CRC Press
Taylor & Francis Group
Boca Raton London New York

CRC Press is an imprint of the
Taylor & Francis Group, an **informa** business

First edition published 2022
by CRC Press
6000 Broken Sound Parkway NW, Suite 300, Boca Raton, FL 33487-2742

and by CRC Press
4 Park Square, Milton Park, Abingdon, Oxon, OX14 4RN

CRC Press is an imprint of Taylor & Francis Group, LLC

ISBN: 978-1-032-21452-8 (hbk)
ISBN: 978-1-032-21454-2 (pbk)
ISBN: 978-1-003-26848-2 (ebk)

DOI: 10.1201/9781003268482

Typeset in Times
by codeMantra

Contents

Preface   vii
Authors   xi

1 Introduction   **1**
  1.1   Internet to Internet of Things   1
  1.2   ICT Standardization   3
  1.3   Convergence   5
  1.4   Industry 4.0 Standards   9
  1.5   Security Issues and Challenges   13
  1.6   Summary   16
  References   17

2 Authorization and Access Control   **19**
  2.1   Introduction   19
  2.2   Threats and Attacks Modeling   20
  2.3   Overview of Authentication and Authorization   24
  2.4   Access Control Paradigms   25
  2.5   Implementation Perspective   29
  2.6   Summary   29
  References   30

3 Open Authorization 2.0   **33**
  3.1   Introduction   33
       3.1.1   OAuth Roles/Main Actors of OAuth2.0   35
  3.2   Motivation   36
  3.3   Protocol Overview   37
  3.4   Use Case   39
       3.4.1   User Agent as Use Case   39
            3.4.1.1   Educational Application   39
       3.4.2   Web Server in Web Application   40
  3.5   Authorization Process   41
       3.5.1   Authorization Code Grant   42
            3.5.1.1   Authorization Code   42
       3.5.2   Implicit Grant   44
       3.5.3   Resource Owner Password Credential Grant   46

	3.5.4	Client Credentials Grant	47
		3.5.4.1 Types of Token	48
3.6	Security Analysis	48
	3.6.1	Phishing Attacks	49
	3.6.2	Countermeasures	50
	3.6.3	Clickjacking	50
3.7	Summary	51
References		51

4	**User-Managed Access**	**53**
4.1	Introduction	53
	4.1.1	Roles of UMA Protocol	55
		4.1.1.1 Resource Owner	55
		4.1.1.2 Client Application	55
		4.1.1.3 Authorization Server	55
		4.1.1.4 Resource Server	55
		4.1.1.5 Requesting Party	55
4.2	Motivation	56
4.3	Protocol Overview	56
4.4	Use Cases	60
	4.4.1	Healthcare Application	60
	4.4.2	Personal Loan Approval Scenario	62
4.5	Authorization Process	62
	4.5.1	Claim Collection	63
	4.5.2	Authorization Result Determination	65
4.6	Security Analysis	65
	4.6.1	PCT and RPT Vulnerability	66
	4.6.2	Cross-Site Request Forgery Attack (CSRF)	67
4.7	Summary	68
References		69

5	**Conclusions**	**71**

| *Index* | | 73 |

Preface

The mind acts like an enemy for those who do not control it.

—Bhagwad Gita

Internet users over the globe have been growing rapidly in the last few years. In the last 12 months, it grew up by 319 million, and in total, it is 4.66 billion over the world. More than 380 websites are created in every minute and are placed on different domains; therefore, the role of security in every aspect over network becomes more challenging these days. Web applications require robust frameworks to provide more security. Vulnerability becomes a key challenge to security these days. In the context of a large number of web applications and to sign up those with required details, a user takes more time. To maintain a user's account credential is also one of the challenges to users as well as cloud platform services dealing with web applications.

In view of this, Open Authorization 2.0 (OAuth 2.0) protocol and extended version of it as a user-managed access (UMA) 2.0 provide faster access control through the smarter token-exchanging method. The main focus of the book is that how we can use OAuth 2.0 and UMA 2.0 frameworks for redirecting/connecting to third-party applications with minimal user credentials that are provided by authorization server or many times resource server. Eventually, users can maintain their number of accounts on various web applications, portals or any social sites only by maintaining single primary account.

Making the use of OAuth 2.0, UMA and strong access control frameworks, we can model robust systems. With this context, this book focuses on various authorization and access control techniques, threats and attack modeling. This book provides an overview of OAuth 2.0 framework along with UMA and security analysis. In the view of cyber-security threats, it is important to provide secure access to data with respect to various web applications. This book also covers issues and challenges with this context.

OAuth 2.0 and UMA frameworks provide more reliability to the user as well as third parties by providing access to it through a token system. The main aim of underlined book is to provide details of authorization and access control with OAuth 2.0 and UMA. Important key concepts are how to provide login credentials with restricted access to third parties with primary account as a resource server. Detailed protocol overview and authorization process along with security analysis of OAuth 2.0 are discussed in this book. An overview of UMA is provided, and how it is differed from OAuth 2.0 is explained in the next part of the book. Finally, this book concludes with research openings and some use cases in it. This book also includes case studies of websites for vulnerability issues. The main contributions of the books are as follows:

- Helpful for user to understand the process of authorization and access control
- Users can understand the significance of authorization and access control in security aspect
- Provides an overview of security challenges of IoT and mitigation techniques to overcome the same
- Provides detailed overview and workflow for Open Authorization (OAuth 2.0) framework/protocol
- Provides detailed overview and workflow for UMA 2.0 framework/protocol
- Helps user to understand how to provide access to third-party web applications through resource server by use of secured and reliable OAuth 2.0 framework
- Provides behavioral analysis of threats and attacks using User Managed Access (UML) base modeling.

Concisely, this book shows how the cyber security plays vital role over a network while connecting web applications to each other and also in sharing confidential information by the use of standard authorization and access control process. This book is encouraging the use of OAuth 2.0 and UMA 2.0 protocols for smooth authorization between resource server and third parties by using requesting party token (RPT), request token (RT) and persisted claims token (PCT). Eventually, by using these standard protocols/frameworks, time complexities at user end decline toward the ground. This book also contributes to focusing on security issues and challenges of IoT which is rapidly spreading globally in all areas.

The main characteristics of this book are as follows:

- A concise and summarized description of all the topics
- Use case and scenarios-based descriptions
- Numerous examples, technical descriptions and real-world scenarios
- Simple and easy language so that it can be useful to a wide range of stakeholders from a layman to educate users, villages to metros and national to global levels.

IoT, Internet applications and security are now fundamental to all undergraduate engineering courses in Computer Science, Computer Engineering, Information Technology, and Electronics and Telecommunication. Because of this, this book is useful to all undergraduate students of these courses for project development and product design in access control, authorization and providing security to underlined use cases. This book is also useful to a wider range of researchers and design engineers who are concerned with exploring critical emerging use cases. Essentially, this book is most useful to all entrepreneurs who are interested to start their start-ups in the field of security, IoT and related product development. This book is useful for undergraduates, postgraduates, industry, researchers and research scholars in ICT, and we are sure that this book will be well received by all stakeholders.

Parikshit N. Mahalle
Shashikant S. Bhong
Gitanjali R. Shinde

Authors

Parikshit N. Mahalle, PhD, earned a BE degree in computer science and engineering at Sant Gadge Baba Amravati University, Amravati, India, and an ME degree in computer engineering at Savitribai Phule Pune University, Pune, India. He earned a PhD in computer science and engineering with a specialization in wireless communication at Aalborg University, Aalborg, Denmark. He was a post-doc researcher at CMI, Aalborg University, Copenhagen, Denmark. He was a professor and the head of the Department of Computer Engineering at STES's Smt. Kashibai Navale College of Engineering, Pune, India. Currently, he is a professor and head of the Department of Artificial intelligence and Data Science at Vishwakarma Institute of Information Technology, Pune, India. He has more than 20 years of teaching and research experience. He is a senior member of IEEE, ACM member, life member of CSI, and a life member of ISTE. Also, he is a member of *IEEE Transactions on Information Forensics and Security* and *IEEE Internet of Things Journal.* He is a reviewer for *IGI Global – International Journal of Rough Sets and Data Analysis (IJRSDA)*, and Associate Editor for *IGI Global – International Journal of Synthetic Emotions (IJSE)* and *Interscience International Journal of Grid and Utility Computing (IJGUC)*. He is a member of the editorial review board for *IGI Global – International Journal of Ambient Computing and Intelligence (IJACI)*. He has published more than 150 research publications, with 1711 citations and an H index of 18. He has published five edited books by Springer and CRC Press. He has seven patents to his credit. He has worked as a chairman of various boards of studies.

Shashikant S. Bhong has more than 7 years of experience and is presently an SPPU-approved assistant professor in the Department of Computer Engineering, Smt. Kashibai Navale College of Engineering, Pune. He earned an ME in computer engineering at Savitribai Phule Pune University, Pune, India, and a BE in computer engineering at Savitribai Phule Pune University, Pune, India. He has published four papers at national and international conferences and journals. He has worked as an assistant professor in STES, Rwanda Kigali, Rwanda (East Africa) in 2016 and as an instructor/trainer in Combat Training Centre (CTC) Gabiro, for the Rwandan army.

Gitanjali R. Shinde, PhD, has over 13 years of experience and is presently an assistant professor in the Department of Computer Engineering at Vishwakarma Institute of Information Technology, Pune, India. She earned a PhD in wireless communication from CMI, Aalborg University, Copenhagen, Denmark, on the research problem statement "Cluster Framework for Internet of People, Things and Services". She earned an ME in computer engineering at the University of Pune, Pune, India, in 2012 and a BE in computer engineering at the University of Pune, Pune, India, in 2006. She has received research funding for the project Lightweight Group Authentication for IoT by SPPU, Pune. She has presented a research article at the World Wireless Research Forum (WWRF) meeting, Beijing, China. She has published 50+ papers at national and international conferences and journals. She is the author of five books with Springer and CRC Press/Taylor & Francis Group. She is also the editor of books with De Gruyter and Springer. She is a reviewer of prominent IGI journal publications and IEEE Transactions.

Introduction

1

1.1 INTERNET TO INTERNET OF THINGS

Science fiction (sci-fi) movies are rapidly turning into a pragmatist or reality. These sci-fi movies are becoming reality due to the Internet. The Internet, also known as "network of network", is one of the most important and transformative powers of technology, which forms the backbone of virtual communication today. It is like a digital gum that is attached to humanity in one way or another, from video calling someone to searching information, from finding location to watching movies at Over The Top (OTT) platform, and there is no area left untouched. Internet is a more straightforward term, a technology that sets up a link between your PC/computer and someone else's PC/computer worldwide through some server, router and switches. The significant type of current Internet communication is among people/human (i.e. human to human). We can call it the Internet of People. The Internet of People changes the world. Well, there's a new Internet emerging and it's poised to change the world you see; this new Internet is not just about connecting people but about connecting things and so it's named the Internet of Things (IoT).

It is a big challenge to connect things to the Internet. Here's why: because things can start to communicate/share their knowledge/experience with other things. Here, the question that bugs our mind is: how is it possible that two different things can communicate or share their experience with each other? This is how it goes: you take things and add some features to the things like ability to sense, communicate, control and touch. By adding this feature, there you create opportunities for the things to communicate, interact and collaborate/team up with other things. This is the same as how humans communicate, interact and collaborate with each other in their own environment with the help of five sense organs by which humans can sense, see/watch, smell, hear and taste. If humans don't have these five sensory organs, then they could also be normal objects or things. Due to this ability to connect or communicate, the IoT and Internet of People intersect. For example, the smartphone used by us is also a thing that has the capability to sense more than we can imagine. It can sense your location, your movement and position of your phone (auto rotation and switch off screen while calling and phone is near to your ear), adjust the brightness of the display screen based on the surrounding light, etc. One can even say that it has the eyes to see in the form of a camera, has the ability to speak using speakers and capability to listen and record via microphone and, most importantly, has the ability to communicate with different devices or phones wirelessly. This is all due to adding some features and sensors/devices to phones like motion sensor, GPS, speaker, mic, ambient light sensor, environmental sensor, proximity sensor, accelerometer sensor, barometer sensor, gyroscope sensor, etc. How about this example, a bracelet which can track steps, all the activity you do, how well you have taken your nap, track heartbeat and pulse rate, and also can communicate. We named this type of thing as a smart band. By just adding some features to this bracelet, we can make the bracelet (thing) to communicate. How about the dog collar? Before the emergence of Internet and IoT, it was just a dog collar, but after that, the collar has become smart: it can track the dog's activity, location, etc. There are hundreds of thousands of such things which become smarter after IoT, just by adding some IoT sensors and devices. The good thing is that we have systems and some tools by which we can add this feature to the already existing things by which they can communicate with each other, and this is all due to the IoT.

The future Internet will be having the ability to connect and communicate with all the physical and virtual things which are surrounding us in the existing Internet. The IoT is a dream that involves connection among various physical and virtual things to see how life would change when things, homes, villages and cities become intelligent like humans. In this context, IoT is the fundamental piece of "Future Internet" that directs people, private and public associations, and educational and research institutes to become smart. By using smart things, they can participate or improve their business, data and social cycles by interfacing with themselves and with the surrounding. Thus, smart things are made to sense the data and react accordingly or autonomously to the occurred events without or with any human interaction, and they should self-configure themselves.

IoT is the next upcoming form of Internet communication where devices will communicate with other devices, which is called machine-to-machine (M2M) communication. IoT can be said to establish the ability of communication among everybody and everything. IoT empowers us to implant a type of intelligence in the objects/things that are or can be connected with the Internet to share data, make communication, give response based on inputs, take decisions independently, and provide all the useful services [1].

1.2 ICT STANDARDIZATION

Information and Communication Technology (ICT) is a more extensive term of Information Technology (IT) which uses Communication Technology. Internet, mobile/cell phones, wireless networks, computers, middleware, software, social networks, and other media applications and services associated with this technology allow to store, retrieve, transfer and manipulate information or data in the digital form. ICT is aimed to work on the seriousness of industry and to satisfy the needs of the general public and economy. India has played an important role for the globalization, development and research of ICT. The reason behind this is, India is developing day by day and there is a massive growth in the field of ICT industrial development and research. Besides India, large-scale research is going on in countries of North America, Asia and Europe in the field of ICT along with research in the advanced wireless IT industry and all the markets associated with it. Parallel to this, the national and international organizations such as the European Telecommunications Standardization Institute (ETSI) [2], the International Telecommunication Union (ITU-R) [3], the Association of Radio Industries and Businesses (ARIB) [4] and the Telecommunications Industry Association (TIA) [5]) are effectively organizing the rapidly growing ICT industry by forming and following specific guidelines and standards.

The Global ICT Standardization Forum for India (GISFI) [6] is an Indian standardization forum which is effectively working in the area of ICT sector and its associated application fields such as telemedicine, energy, biotechnology and wireless robots. With ICT and its products, applications and services, services and applications are growing quickly in the present worldwide economy turning into a vital piece of our regular daily existence, and it is of basic significance to establish an environment that addresses both the overall industrial business and society's expectation. To increase the intensity of the industrial business while guaranteeing all the citizens or human society that can benefit from the opportunity created by the advancement of research in the field of ICT. Like other standardization forums, GISFI focuses on balancing between India and worldwide industrial requirements with India and other countries' society expectations.

ICT is a significant driver of seriousness and representation, and right now, it is one of the key modern industrial sectors. Standardization of ICT is very important for today's world to provide a strong framework for growing the economy and development in the society. Standardization is the voluntary help shared among industries, public organizations, private organizations, consumers and all the other interested parties for development of technical specifications to compete in the global market. In today's competitive world, industries use standardization for growing, which ultimately leads to market growth.

The significance of ICT standardization for advanced digital inclusion has been featured in various events and discussion forums. GISFI [6] characterized standardization as one of the fundamental structure blocks of the Information Society. In fact, standardization is not only essential for ICT field, but it is also necessary to maintain a stable economy [7]. The standards and rules to manufacture or develop a new product/technology and services are accounted for in the economy of distribution and production, by keeping the cost of product low and facilitating access to worldwide markets, which can improve the economy of an organization or an industry. Standardization of ICT provides a platform for organizations of ICT to distribute their standard products and services, and compete in global markets. Due to standardization of ICT, competition in the world market increases, which leads to a lot of research and development in ICT.

There is one more fact that there are many developing countries which are lagging in research and development of digitalization of ICT due to their financial conditions. These countries have no option left; hence, they implement the ICT developed by the developed countries. Now the question is, will the ICT technology made in a foreign or developed country (let's say X is the developed country) be suitable for the already existing infrastructure of a developing country (let's say Y is the developing country)? In other words, can X technology be implemented in Y country's infrastructure? Or how can Y country improve their infrastructure so that they can use X technology of ICT? And the answers for these questions is one; i.e. before developing a new ICT technology, a discussion should be made with the stakeholders or the ICT standard forum of different countries to develop or design some standards for that technology.

There are some innovations in ICT like Wi-Fi, i.e. Wireless Local Area Network (WLAN), a communication protocol based on IEEE 802.11 standard. Wi-Fi is used by almost all the countries worldwide. For devices/computers to connect and communicate wirelessly, computer and wireless communication device manufacturers follow some industrial standards, thus providing uniform wireless connectivity between devices. Another example is Next-Generation Network (NGN). Development of NGN standardization allows voice, video and data services to be transferred by convergence based on Internet protocols. The standardization in ICT has solved many global issues and problems by avoiding differences between developing and developed countries. Figure 1.1 shows research and development in the field of ICT standardization from 1998 to 2008 [7].

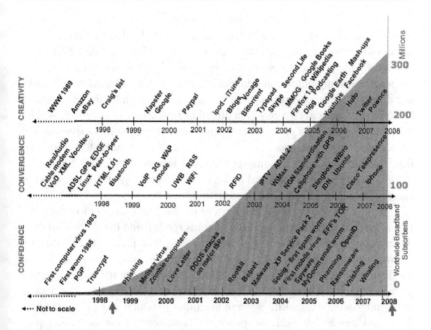

FIGURE 1.1 Technology developed in the field of ICT

The convergence of networks from different network services from different parts of the world is standardized and transformed into a uniform global network by developing specific guidelines, thus enabling smooth access to a wide variety of Internet services across the world. Thus, standardization has helped to speed up the application, acquire new technology or service of ICT, and direct market production and sale, thus providing a strong support to drive the ICT industry.

1.3 CONVERGENCE

The IoT has given immense benefit to all the industries with all possibilities of automation, catching new bits of knowledge and keeping individuals and organizations more secure. In today's world, there are many new technologies being developed which need to integrate with the IoT to compute different problems and thus make people's life easier. When two or more technologies come together and form a new technology to solve a problem in human life, it is called convergence. When IoT and some other technologies come together to form a whole new technology that solves issues and problems in an industry or in day-to-day life, it is convergence of IoT. In this section, we will discuss different technologies that are converged with IoT to solve issues and problems. One good example is converging wireless and wired networks for proper functioning of telecommunication. Both wired and wireless networks use different technologies and Internet protocols to communicate; hence, to be converged, they need a common digital infrastructure or platform. Before, broadband service was provided through optical fiber cables, but from the last decade, due to the emergence of WLAN, we are able to connect to the Internet from IoT devices such as smartphones and laptops wirelessly. Figure 1.2 shows the convergence of wired and wireless technology in the direction of common digital platforms or infrastructure [6].

As we know, IoT sensors and devices produce huge amounts of data. So, to use this data efficiently, we have to converge IoT and big data to analyze the product by IoT devices and sensors. For example, in the electricity sector, to avoid energy wastage, we need to converge IoT sensors and electrical devices for gathering information and data regarding usage and alerting people to avoid wasting energy. Individual devices will collect the data in real time from houses, buildings and machines to which they are connected. This data can be cleaned and processed, and sent to the electricity provider who would analyze the data and supply energy efficiently, which will decrease the loss of electricity and improve the life of people. In Ref. [7], the author developed an agent collecting information regarding energy usage for smart houses and buildings using IoT and big data. There are many energy management systems that are developed to control and manage the use of energy. But in Ref. [7], the authors use a hybrid network by combining IoT

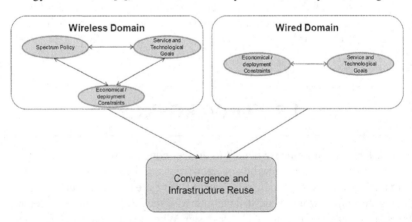

FIGURE 1.2 Convergences of wired and wireless technologies in the direction of common digital platform or infrastructure

and big data to manage and control the use of energy efficiently. In Ref. [8], the authors developed a framework based on weather detection and sensor fault detection using IoT and big data, where the role of IoT is to collect the data, and using big data techniques, this data is pre-processed and then some meaningful information is taken out from the data. The authors have used k-means, KNN and some other machine learning techniques to analyze the weather and detect faulted sensors in the network. Another example of its application is in the field of military, where sensors can be implemented in machinery, guns, transport equipment and weapons used by the military, and data are collected based on past trends and analyzed to derive information such as prior maintenance of all equipment. By implementing this technology, the military can maintain its security equipment efficiently. IoT and big data can be used in sports where we can collect data about the health of athletes before, after and during the game. Using this data, we can predict the injuries that can happen to the players, and thus help whole team or the concerned player compete more energetically and efficiently. IoT and big data can also be implemented and used in health, aviation and many other fields where huge amount of data is produced.

To ensure accelerated success in the IoT world, we require services that provide high performance, reliable, accurate and trustworthy computation of data, and this can be provided by cloud technology. By converging IoT and cloud, we can achieve a high computation network as IoT devices are of low power and have low computing capacity. There are two different approaches. The first one is cloud-based IoT which means including IoT functions into IoT, and the next one is IoT-centric cloud which means including cloud into IoT [9]. In Ref. [10], the authors proposed a cloud-based IoT data gathering and processing platform where a platform based on IoT and cloud is used to store the data, as data generated by the IoT devices is huge and needs some place to store and process. A set of local automation clouds were implemented using the Arrowhead framework and support core services in Ref. [11], where the authors enabled IoT automation using local clouds. In Ref. [12], the authors proposed a new approach on sharing IoT devices among end users. Based on a service-oriented approach, IoT devices expose data and action resources that are available within a cloud platform anytime from anywhere. Recently, the Indian government has implemented fast-tag to collect the toll tax. This application is based on IoT device, i.e., RFID tags, and cloud is used to store and maintain the data. In Ref. [13], a model is proposed for automatic toll tax collection using RFID and cloud storage. In Ref. [14], the authors proposed a prototype based on e-health using IoT and cloud computing, for example, a smart band tied on the wrist can measure your health-related data such as the number of steps while walking and pulse rate. This data can be stored on the cloud and can give you alerts about your health when necessary. In Ref. [15], the authors proposed by smart dustbins for metro stations and Delhi city in which IoT sensors are used to check if dustbin is full or not and if garbage is properly thrown in wet or dry dustbin. All this data is stored in cloud, and when a dustbin becomes full, it alerts the respective authority to empty the dustbin. A smart parking system is proposed in Ref. [16], where the authors used cloud systems and Raspberry Pi. The application was implemented in the SNU University, where with the help of Raspberry Pi and IoT sensor, they captured the information of parking and send it to cloud systems from where the whole parking system is managed. The above-discussed technology and applications are some examples of IoT and cloud conversion which make human life easy.

IoT technology can also converge with Artificial Intelligence (AI), machine learning and deep learning. The data generated by IoT devices can be used to predict, automate or detect faults in a system/application. For example, diagnosis of crop diseases using image-based deep learning mechanism was proposed in Ref. [17], where with the help of sensors, images of strawberry plants were captured and the data/images were sent to an application based on convocation model that predicts the condition of plants or diagnoses the plant disease. In Ref. [18], an intelligent home access control system using deep neural networks was proposed in which with the help of sensors and camera, they send packet to their system based on deep learning and provide access to the person. In Ref. [19], the authors proposed a framework to recognize human activity in smart homes using deep learning algorithms. This system collects data of human behavior in a house, and using deep learning, these activities are recognized and users in the house are alerted about their activity for activity management. In the health system, IoT and AI play very important roles nowadays. Many health-related machines are manufactured based on IoT to collect patient data to be used in the diagnosis of diseases. In COVID-19 pandemic, lots of research and development have been done on developing applications. For example, IoT-based tool-generated X-ray images are analyzed by deep learning, machine learning or AI to find out the presence and severity level of COVID infection in patients. In the manufacturing industry where big machinery is used, we can use IoT sensor and AI to analyze the machinery for maintenance purpose. There are many such fields and applications where IoT and machine/AI/deep learning can be used to increase automation and efficiency of systems.

As the IoT system is converging with many technologies, lots of data are generated and transferred between IoT devices. Therefore, to secure the communication between IoT devices, there should be convergence between IoT and security technology. In Ref. [20], the authors designed a framework for identity and trust management and access control using IoT devices and fuzzy approach. In Refs. [20,21], the authors proposed a Base method to calculate trust in device-to-device communication in IoT using fuzzy approach and machine learning model, and proposed a model that can be used to provide security in device-to-device communication in IoT.

Hence, convergence of technologies gives great and innovative technologies to the world, which makes human life simple and automated, while providing a platform to industries/organizations to predict the future of each and everything using which they can make effective use of data to make a profit.

1.4 INDUSTRY 4.0 STANDARDS

Let's go back around 10,000 years when our ancestors used to collect food through foraging and wandered around for collecting food. They were eating fruits, vegetables or whatever they could get or collect. Then, it transferred to farming where basically varied crops, plants, vegetables and fruits were grown. The transformation from foraging to farming resulted in an increase in production, communication between humans, transportation and much more. So, as the population increased, the production grew as well. Then came the Industrial Revolution where new technologies and machines were produced and new approaches to the production were introduced. This basically shifted the simple economic models to more aggressive economic models and social architecture. So basically, the Industrial Revolution went through different stages. The period from 1760 to 1840 was the First Industrial Revolution where invention of steam engine and construction of railway stimulated the overall revolution which resulted in utilization of machines in production. Then came the Second Industrial Revolution during the transition from the 19th century to the 20th century where electricity and assembly lines triggered the revolution which resulted in mass production machinery that could use electricity for faster production. Then, in the 1960s, when computer was starting to get popular, naturally there was increase in different computer and computing devices, peripherals, etc., and the transformation in the industries also happened. There was increased use of digital technology in the industries, so the use of computer and digitalization was another revolution. The Third Industrial Revolution again increased the production of goods and commodities in the industries. This production was due to increase in the use of semiconductor and semiconducting devices and that was almost in parallel with the growth of computers. The result of this revolution was an increase in computing technology such as mainframe computers, personal computers and the Internet (connecting all these computers and computing devices).

The Fourth Industrial Revolution, i.e., Industry 4.0, originated in the German economy. It was started as individual industries having their own individual IT infrastructure through the Third Industrial Revolution, but the questions arise here are how could we improve the production even faster? And how could we make processes even more efficient? So, researchers thought about how things could be done or how we can make processes even faster and more efficient. This happened because there were different sensors and sensing technologies which were developing in parallel and becoming popular. Introduction of sensors, actuators, etc., along with regular IT infrastructure and the Internet, was able to transform the existing IT-based infrastructure companies to much more efficient ones to connected sensed machinery. This is the Fourth Industrial Revolution or Industry 4.0 that we are going through at this moment. During the revolution, sensors became cheaper and more powerful, and their sizes became smaller. Extensive use of technology involving AI, machine learning, cyber security, IoT etc. is witnessed in Industry 4.0. Figure 1.3 shows features of Industry 4.0 technology and its contributions toward digitalization.

Research and development in Industry 4.0 has created many opportunities in smart-connected machines, smart factories, smart house/city, gene sequencing, nanotechnology, renewable energy and quantum computing.

Key component in this revolution is the IoT, a network of smart devices facilitating exchange, transfer, and analysis of data while driving efficiency

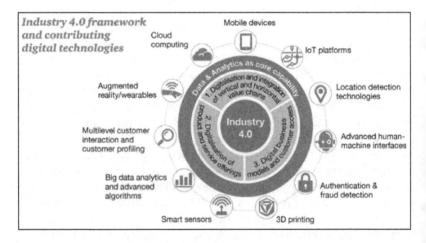

FIGURE 1.3   Industry 4.0 technology features and contributions toward digitalization (Source: Industry 4.0: Building the digital enterprise, 2016 global Industry 4.0 survey, PwC engineering, & construction 2016.)

and innovation across every aspect of our society. Standardization of this continually evolving ecosystem is a complex task. To make or present some guidelines and standards for Industry 4.0 is a complex task considering all the challenges and opportunities in this agile, all collaborative world. Also, data privacy, reliability and security take a new issue in this interconnected world. Although all the standards are mostly extended or adapted by the different industries, this standardization creates one common understanding between industries. So, many different organizations have created a reference standard which is aligned with Industry 4.0.

The Deutsches Institut für Normung (DIN) – German Institute for Standardization with many different organizations and stakeholders – has published the research on topic "Reference Architecture Model for Industry 4.0 (RAMI 4.0)" [22]. The Reference Architecture Model describes the fundamentals of Industry 4.0. Figure 1.4 shows layers of RAMI model and connection between IT sector, manufacturing plant/industry and product life cycle in 3D space. The vertical axis of the left side represents the IT sector's perspective, ranging from physical device (asset) to complex function and mapping to business model. The horizontal axis at left side indicates product life cycle; type and instance are distinguished as two

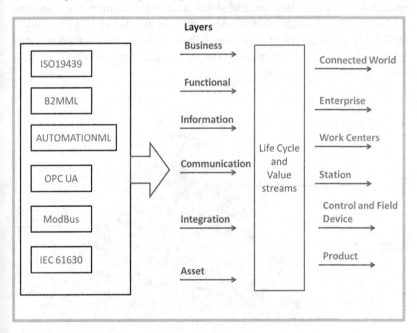

FIGURE 1.4   RAMI model

main concepts. On the right side, the horizontal axis shows location of responsibility and feature in hierarchy. RAMI represents an architecture where product is the lower level and competitive world is at the top level. Let's take the example of Open Platform Communications (OPC) UA in the IT communication layer where OPC UA establishes M2M communication on some stable/constant data format which helps in data analysis and management processes.

The research on "Standards Landscape for Smart Manufacturing Systems" was published by the National Institute of Standards and Technology (NIST) – a US organization [22]. In this research, first, standards are classified with their respective function. ISA95 Reference model leveled in pyramid structure where at lower level devices are kept and at higher level there is enterprise (shown in Figure 1.5). Second, standards were organized taking into account different phases of the product life cycle such as modeling practice, data exchange and product model, product category data, manufacturing modeling data and data management. Finally, the standards were classified on the basis of the production system life cycle. In this phase, they included more categories related to production such as production system engineering, production system model data, and production system maintenance and operation. The ISA95 Reference Model is classified standard in more general than RAMI model and also this model is not multidimensional. In this model, OPC UA is classified in SCADA which focuses on the information or data, but not on how to use the information. As in RAMI model, OPC UA was focusing on data and its use.

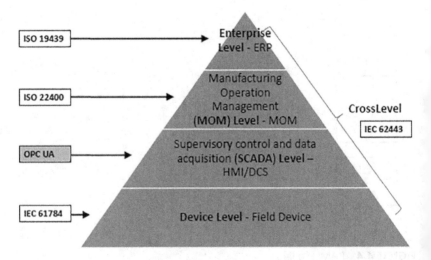

FIGURE 1.5 ISA95 Reference Model [22]

The Ministry of Industry and Information Technology of China with Standardization Admiration has submitted research defining "National Smart Manufacturing Standards Architecture". They have also classified architecture into three dimensions as follows: (1) life cycle: from design to production to logistics to marketing to sales and service; (2) smart functions: from resource element to new business models; (3) hierarchy levels: from device to inter-enterprise collaboration. Their aim was to classify the model with respect to function.

Standardization can likewise assist you in creating business flexibility expected to change your business under Industry 4.0 or according to the difficulties delivered by COVID-19. It is clear that organizations/industries that took on digital solutions are better positioned to effectively face any emergency (like COVID-19 pandemic). Using standard tools of Industry 4.0, anyone can grow their business in difficult situations as well as in the new normal world.

1.5 SECURITY ISSUES AND CHALLENGES

In Industry 4.0, the role of IoT devices has increased. IoT devices are used in almost every field available in the world. Due to the massive use of IoT devices and Internet, lots of personal and public data are transferred in huge network of devices. So in IoT, there are many issues and challenges related to design, security and privacy. Previously, there were many attacks on IoT devices; some famous attacks are "The Dallas Siren Attack" – 156 emergency siren went off at around midnight leading to panic; "The Teddy Who Was Spy" – connected teddy bears leaked 2 million parents and kids message recording; and "Hackers Kill a Jeep Remotely" – hackers took control of the vehicle at 70 mph and the driver lost total control over the vehicle. Many other data breaches have also happened in the past where information of credit cards and accounts were leaked.

Key Challenge Areas
  1. **Access Control and Authentication**: Access control and authentication is one of the most important key issues for security and privacy. Access control means which type of access is given to different types of user/devices. Authentication check is if the user/ device is authenticated to access the IoT device.
  2. **Confidentiality**: Protecting the information from unauthorized users/devices. We can allow the user who is authorized to provide access to different users for sensitive information.

3. **Integrity**: It is overall accuracy, completeness and consistency of the data. That means data should be changed during the transmission and steps to be taken to protect the data from unauthorized access.

4. **Availability**: The information or devices should be available at the time when needed, or else it will be useless.

5. **Privacy**: It is a right of devices/users to have control of when and how their sensitive data is collected or processed in the IoT environment.

6. **Trust**: Calculation of trust on one user to another is difficult, and it's still a big challenge for security in IoT.

7. **Identity Management**: Managing identity of users and devices is a challenge, and IoT falls short in managing digital identity.

Security Challenges and Issues in IoT

1. **Lack of Encryption**: Encryption is one of the ways to secure the data from hackers from accessing the data and leading to IoT security challenges. IoT devices are lacking in storage and performance than the traditional computer system due to the size and need of IoT devices. So, due to restriction in size and performance, we need to use low-power cryptography algorithms to provide security. This results in an increase in the possibility of attacks as the hackers can easily manipulate the low-power cryptography algorithms.

2. **Insufficient Testing and Updating**: With increase in the demand of IoT devices, manufactures are more engaged in producing devices as fast as possible without giving too much importance to the security. As most of these IoT devices and products do not get time for testing and updating, this is an advantage for hackers to hack and exploit other security issues.

3. **Brute Forcing and Issue of Default Password**: Weak password and login details or credentials can lead to vulnerability to password hacking and brute forcing. Any organization that uses factory default login details on their devices is placing their business, assets and the information of their valuable customers at risk.

4. **IoT Malware and Ransomware**: As the number of devices increases, ransomware uses encryption to effectively lock users from several devices and platforms and can steal user valuable information. For example, there are CCTV cameras used in organizations. Hackers can hijack the CCTV camera and take pictures and videos, and can also turn off the camera. By using malware

access point, hackers can demand ransom money or information to unlock the devices and return the data.

5. **IoT Botnets Aiming at Cryptocurrency**: IoT botnet workers can manipulate the data privacy, which could be risky for the crypto market. The Blockchain is trying to boost security. Blockchain technology is not vulnerable, but application development could be.

Design Challenges in IoT

1. **Low Battery Life**: Battery life of IoT is limited as IoT devices are small in size.
2. **Increased Cost and Time to Market**: Embedded systems are tightly constrained by the cost, so the cost of the devices should be low. Designers also need to solve the design time problem and bring embedded systems on time to market. So, high function should be provided in low cost by the designing team.
3. **Security of the System**: System has to be designed robustly and reliably, and should be secure with some cryptographic algorithm and security protocol. We can have convergence of IoT and some cryptographic or security technology, but security technology should be feasible or light weight, so that it can be implemented in our low-configured IoT devices.

Development Challenges in IoT

1. **Connectivity**: IoT devices should be feasible to connect to different technologies like cloud, big data, AI, etc. Connected devices that provide the information are extremely valuable. But poor connectivity can become a challenge where IoT sensors are used to monitor and transmit the information.
2. **Gross Platform Connectivity**: IoT devices should be developed by keeping in mind technology changes in the future. Development is required to balance between hardware and software. It is a challenge for the developer that the IoT platform should deliver the best performance – not heavy OS, device update and bug fixing.
3. **Data Collection and Processing**: We know IoT data plays an important role, but the important thing here is processing and storing the data. Along with security and privacy, developers should ensure that they plan well for storing and collecting the data and also processing the data within the environment.
4. **Lack of Skill Set**: All the developmental challenges can only be handled in a proper way if there are experienced skill set developers working on IoT applications.

To sum up the issues and challenges in IoT, it is required to be smart enough to make a choice of IoT device based on its safety and security capacities. While planning, designing or developing a secure IoT system/framework, it is important to go through the issues of IoT devices, by thinking about network infrastructure, vulnerability and organizational risk.

1.6 SUMMARY

Internet and IoT support a wide range of cool applications from sustainable development (smart city) to movement/transportation (self-driven car), from healthcare to agriculture and manufacturing and packaging. Some of these applications are smart waste management, smart street light, smart street parking, collective vehicle, sleep monitoring, enhanced adherence – ingestible sensor, precision agriculture, food safety and storage efficiently, smart manufacturing, etc. This technology or application of IoT can be connected anywhere in a network. So IoT is the "next big thing" and will be having ten times more users in future than those currently using mobile Internet. IoT is the combination of sensor, security, circuit and, most importantly, Internet which leads to creation of cool applications.

The first section describes how the Internet has grown to IoT or Future of Internet is IoT and what are some challenges to connect the things with the Internet. The second section is all about information and telecommunication technology and its standardization, how ICT and its standardization will help IoT to grow more in both technology and economical perspective, and also about many different organizations coming to one platform and setting the standard for ICT which will help to develop both developing and developed countries. The third section is about convergence of IoT with different technologies, and there is a discussion about some examples of cool applications and technology-based convergence of IoT. The fourth section is all about Industrial Revolution Four (or Industry 4.0), and there is a detailed description of all the four revolutions or industry shift and Industry 4.0 standard and reference model. There is some discussion about how this standard will help us to grow technologically and economically and how Industry 4.0 technology will grow the economy of the world. In the last and fifth section, there is a discussion on some key challenges and issues faced by IoT developers while designing and developing IoT devices or system or platform or framework.

REFERENCES

1. K. Chopra, K. Gupta and A. Lambora. (2019). Future internet: The internet of things-a literature review. In *2019 International Conference on Machine Learning, Big Data, Cloud and Parallel Computing (COMITCon)*, pp. 135–139, doi: 10.1109/COMITCon.2019.8862269.
2. European Telecommunications Standards Institute (ETSI), at www.etsi.org.
3. International Telecommunications Union, ITU, at www.itu.int.
4. Association of Radio Industries and Businesses (ARIB), at www.arib.or.jp/
5. Telecommunications Industry Association, at www.tiaonline.org.
6. Global Standardization Forum for India (GISFI), at www.gisfi.org.
7. Canazza, M. (2009, August). Global effort on bridging the digital divide and the role of ICT standardization. In *2009 ITU-T Kaleidoscope: Innovations for Digital Inclusions* (pp. 1–7). IEEE.
8. A. C. Onal, O. Berat Sezer, M. Ozbayoglu and E. Dogdu. (2017). Weather data analysis and sensor fault detection using an extended IoT framework with semantics, big data, and machine learning. In *2017 IEEE International Conference on Big Data (Big Data)*, pp. 2037–2046, doi: 10.1109/BigData.2017.8258150.
9. A. R. Biswas and R. Giaffreda. (2014). IoT and cloud convergence: Opportunities and challenges. In *2014 IEEE World Forum on Internet of Things (WF-IoT)*, pp. 375–376, doi: 10.1109/WF-IoT.2014.6803194.
10. H. Park, E. JeeSook and S.-H. Kim. (2018) Crop diseases diagnosing using image-based deep learning mechanism. In *Second International Conference on Computing and Network Communications (CoCoNet'18)*, Astana.
11. R. Xue, L. Wang and J. Chen. (2011). Using the IOT to construct ubiquitous learning environment. In *2011 Second International Conference on Mechanic Automation and Control Engineering*, Inner Mongolia.
12. F. Hongqing and H. Chen. (2014). Recognizing human activity in smart home using deep learning algorithms. In *Proceedings of the 33rd Chinese Control Conference*, July 28–30, 2014, Nanjing, China.
13. A. Javed, H. Larijani and A. Wixted. (2018). Improving energy consumption of a commercial building with IOT and machine learning. In *Published by the IEEE Computer Society* 1520-9202/18/$33.00 ©2018 IEEE.
14. U. S. Shanthamallu, A. Spanias, C. Tepedelenlioglu and M. Stanley. (2017). A brief survey of machine learning methods and their sensor and IOT application. In *2017 8th International Conference on Information, Intelligence, Systems & Applications (IISA)*, Larnaca.
15. A. Kanaway and A. Sane. (2017). Machine learning for predictive maintenance of Industrial machines using IOT sensor data. In *2017 8th IEEE International Conference on Software Engineering and Service Science (ICSESS)*, Beijing.
16. P. N. Mahalle, P. A. Thakre, N. R. Prasad and R. Prasad. (2013). A fuzzy approach to trust based access control in internet of things. In *Wireless VITAE 2013*.

17. V. C Emeakaroha, N Cafferkey, P. Healy and J. P. Morrison. (2015). A cloud-based IoT data gathering and processing platform. In *3rd International Conference on Future Internet of Things and Cloud*, Rome.

18. R. Biswas and R. Giaffreda. (2014, April) IoT and cloud convergence: Opportunities and challenges. In *Proceedings of IEEE World Forum Internet Things (WF-IoT)*, pp. 375–376, New Orleans.

19. Y. Benazzouz, C. Munilla, O. Günalp, M. Gallissot and L. Gürgen. (2014, March). Sharing user IoT devices in the cloud. In *Proceedings of IEEE World Forum Internet Things (WF-IoT)*, pp. 373–374, New Orleans.

20. P. N. Mahalle, B. Anggorojati, N. R. Prasad and R. Prasad. (2012). Identity authentication and capability based access control (IACAC) for the internet of things. *J. Cyber Security Mobility* 1, 309–348. Doi: 10.13052/jcsm2245-1439.142.

21. R. V. Patil, P. N. Mahalle and G. R. Shinde. (2020). Trust score estimation for device to device communication in internet of thing using fuzzy approach. *Int. J. Inf. Tecnol.* Doi: 10.1007/s41870-020-00530-9.

22. I. Grangel-González, P. Baptista, L. Halilaj, S. Lohmann, M. E. Vidal, C. Mader and S. Auer. (2017). The industry 4.0 standards landscape from a semantic integration perspective. Doi: 10.1109/ETFA.2017.8247584.

Authorization and Access Control

2

2.1 INTRODUCTION

Authentication, access control and authorization mechanisms are essential parts of communication in today's world to provide secure access to users. Users can access thousands of services at anytime from anywhere using a heterogeneous type of network. There are mainly two reasons for the security measures to be mandatory: first, the data of almost all organizations are kept on cloud, and users can access these data using various API/services, and second, due to the emergence of Internet of Things (IoT), communication services can be accessed from any device without restrictions of time and place [1]. Hence, for secure access of services, security mechanisms must be deployed.

A large number of services are deployed for the betterment of users' life [2]. For example, e-health is another important application. Services of this kind of application make life of users easy by defeating the location and time limitations. Services of e-health are categorized based on the role of users in a system, i.e. specialist doctor, intern, nurse and patient. A single service may provide different access rights to different users with different roles, i.e. doctor has different access rights than the nurse for the same service. For example, for examining and updating health records of a patient, doctor is authorized to update the record, but the nurse may not have the same authorization. If the proper authorization and access control mechanisms are not used, then the patient's sensitive information may be misused.

These massive application services makes a user's life complete and easy with full of leisure and comfort. But the other side of the coin is "To get secure access to these services at the verge of the moment".

DOI: 10.1201/9781003268482-2

Security is a combination of five building blocks, i.e. confidentiality, non-repudiation, integrity, authentication and access control. Access control and authorization are vital issues for the secure use of services. Access control mechanism is the process to prevent unauthorized access of system resources and services [3–5]. Access control process starts with the identification of a person/device (authentication) followed by granting/denying role-based access permissions to the person/device in a system/network. Authentication checks whether a person/user is a legitimate one, authorization defines the rights the user is having, and access control denies/grants access to the user.

2.2 THREATS AND ATTACKS MODELING

Data on the cloud is more vulnerable as there can be many threats for data in a network. It is important to understand the various threats to prevent security attacks. In this section, modeling of various threats is presented [6].

- **Cloning of Device**: The clone of an original device is manufactured with the same attributes of the original device. The cloned device can be used by hackers/eavesdroppers to hack the network and steal the data. Sybil and node replication attacks can be performed on the network using cloned devices.
- **Commissioning of a Device**: The security information such as configuration parameters and key attributes can be compromised due to their presence in wireless media at the time of commissioning the things. Using the key attributes, eavesdroppers can retrieve the secrete key and hack the communication.
- **Malicious Replacement of Device**: The attacker can add a new device to the network, and that device can read all the communication in the network. It can steal information by diverting information toward the malicious server. These devices look for a device with minimum security measures and attack such device.
- **Unreliable Communication Link**: The unreliable communication link is the major reason for security attacks. The data integrity can be questionable in such communication. The successful delivery of data is not guaranteed in communication through the unreliable link, and the retransmission of data is done if data packets are not received in time. In such cases, hackers can perform man-in-middle attacks.

- **Privacy Threat**: It may be possible that application services carry user's sensitive data, e.g. health data. In such cases, security attacks can lead to a privacy threat. Leakage of personal data can result in serious risk for the user.
- **Phishing**: In phishing, a person's data such as login information and credit card credentials are stolen by the attacker. This information can be used in various ways which result in economic and social loss.
- **Ransomware**: It is the software that comes along with a legitimate file through email. When a user opens the attachment, this software takes control of the user's resources. It blocks the resources. To regain the access to the resources, the user may be demanded by the hacker to pay ransom, and the hacker keeps ownership of the resources until the amount is paid.

  There are various threats to security while communicating and working through the Internet. These attacks are performed by attackers due to weak security measures deployed in the network. In this session, the behaviors of various attacks are discussed.
- **Sybil Attack**: In this attack, the attacker takes the identity of a legitimate device and makes an influence over the network by creating multiple duplicate identities, e.g. fake reviews on the e-commerce platforms like Amazon. The workflow of the Sybil attack is presented in Figure 2.1.

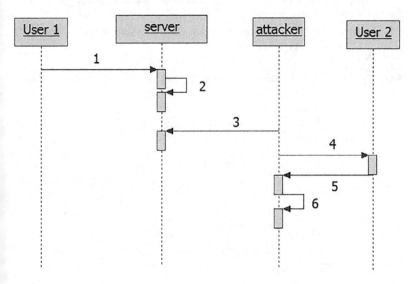

FIGURE 2.1   Sybil attack.

Step 1: User 1 (legitimate) sends authentication request to the server.
Step 2: User 1 gets access to the network after authentication by the server.
Step 3: An attacker gets the identity of user 1 by communicating with the server.
Steps 4 and 5: The attacker starts interacting with other users, i.e. user 2, in the network by pretending that user 1 is communicating.
Step 6: The attacker gets access to other users' data through the identity of user 1.

- **Data Integrity Attack**: Data integrity is compromised in this attack, and then fake data is circulated in the network. The attacker attacks one of the legitimate devices and makes changes in this device, thus making a malicious device. The malicious device modifies the original data received from other users and transmits fake data in the network. The workflow of this attack is shown in Figure 2.2.

Steps 1 and 2: An attacker attacks a network and gets credentials from the devices in the network, i.e. intermediate device.
Steps 3 and 4: Then the attacker embeds malicious code into the device and creates a malicious device with the credentials of intermediate device.
Step 5: The malicious device then alters the data received from destination devices.
Steps 6–8: The malicious device transmits the altered data to source device in the network.

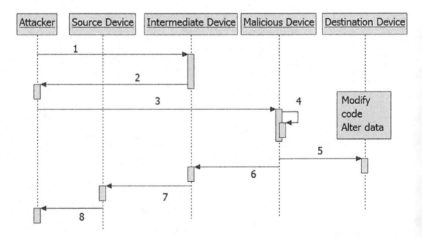

FIGURE 2.2   Data integrity attack

- **Denial of Service Attack**: The aim of this attack is to make service unavailable for intended users. In this, the attacker floods multiple requests to the server and makes it busy. It is difficult for the server to handle a large number of requests, and as a result, the server is not able to answer requests from legitimate users.
- **Man-in-the-Middle (MitM) Attack**: In the MitM attack, public key of the sender (S) is stolen by the attacker, modified by adding the attacker's public key and forwarded on the network. Device R wants to send data to device S, but device R encrypts this data with a modified public key of device S and forwards the data on the network. The data is decrypted by the attacker as it is encrypted by its public key, and the attacker can modify the message and then forward it to device S. Device S will never understand that the message is compromised. The workflow of the MitM attack is presented in Figure 2.3.

Step1: User A requests for the public key of user B for starting the communication.

Step 2: An attacker captures user B's response.

Step 3: The attacker sends its own public key to user A instead of user B's public key.

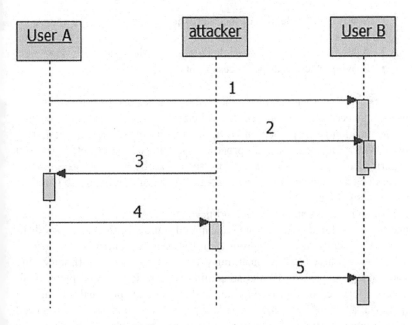

FIGURE 2.3 Man-in-middle attack

Step 4: The attacker captures data from user A, and reads/alters that data and again encrypts the data with the public key of user B.

Step 5: The attacker sends the altered data to user B; hence, both users A and B will never understand that the message is read by some eavesdropper.

2.3 OVERVIEW OF AUTHENTICATION AND AUTHORIZATION

The attacks discussed in the above section are the major concerns for various organizations as their valuable data is present on the cloud and in services provided through the Internet. By means of the service discovery mechanism or various portals, the user will be aware of services available in a specific context. In order to use these services, the user has to ask for service access, and the access should only be granted to authenticated users with proper authorization [7]. Step-by-step processes for providing secure access to the services are as follows:

- Verifying the identity of the user who is requesting for resource/ service
- Allowing/denying access rights to the specific user
- Providing access as per access rights.

The first step is authentication. Authorization that defines the access rights for users and access control enforces the policies mentioned in the authorization. Authentication is the process of verifying identity of the user [8] requesting the service. Authorization is the process of defining policies whether a person is permitted to perform certain actions, such as accessing a resource. Finally, access control mechanisms provide access to the services/resource based on the access policies.

In authentication, a receiver end needs to verify the source of the data and be assured that it came from the claimed source. Authentication is difficult as it typically requires appropriate infrastructure. Authentication is typically done by three factors, something you own (e.g. identity card), something you know (e.g. passwords) and something you are (e.g. retinal pattern). To access a service, the user is requested to disclose some personal information (personal attributes), leading to a risk of privacy violation, so there is a need of a framework for authentication with minimum information disclosure.

Traditionally, for user authentication, different identity management standards have been used such as OpenID [9], Liberty Alliance and CardSpace [10]. Other important initiatives are OpenAM [11], User-Managed Access (UMA) [7] and the Fast Identity Online (FIDO) Alliance [12]. OpenAM is an open standard for authentication based on the JAAS (Java Authentication and Authorization Service) and fine-grained authorization based on XACML (Extensible Authorization Mark-Up Language). FIDO Alliance is a password-less authentication mechanism supported by the Universal Authentication Framework (UAF). FIDO uses a standard challenge-response mechanism for authentication, i.e. public-key cryptography [12].

Nowadays, these are being replaced by a number of technologies and standards, which enable fine-grained control of user attributes and minimum disclosure of personal information [13]. Digital identities should be based on claims [14], which contain an assertion that the user possesses a set of user attributes. Claims are presented to the relying party of the service provider, who evaluates the claim and decides whether to grant access to the service.

In a claims-based identity system, the user makes a claim, i.e. presents a collection of attributes to access resources, the assertion party adds credibility to the claim presented by the user and the relying party evaluates the claims to authenticate the user [14]. The privacy of the user is maintained, as attributes required to access resources may not be enough to reveal the identity of the user. A claim-based credential provides authentication of the user, but for authorization, it is necessary to add an authorization framework like OAuth 2.0 framework [15]. The authorization framework OAuth 2.0 is discussed in Chapter 3.

OpenID Connect was launched in February 2013 by the OpenID Foundation [16]. In this, an identity layer is added on top of the OAuth 2.0 framework. OpenID Connect provides authentication and authorization for single sign-on but faces limitations such as a single point of failure, prone to the attacker, no control over the user list and no way to know how reliable the system is.

2.4 ACCESS CONTROL PARADIGMS

Access control mechanism denies/grants access to the resources based on the access rights. The access control system consists of few components for making access decisions, i.e. Policy Enforcement Point (PEP), Policy Information Point (PIP), Policy Decision Point (PDP), and Policy Administration Point (PAP), as presented in Figure 2.4. PEP is responsible for protecting the resources, user access requests are received by the PEP and it redirects it to

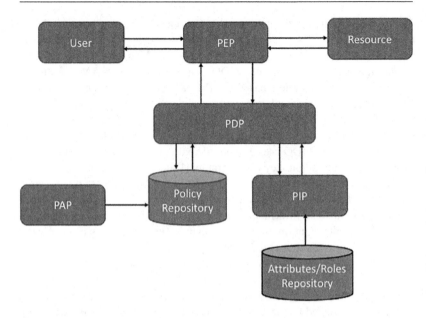

FIGURE 2.4   Access control

the PDP for making access decisions. PDP verifies the user role/attributes and based on these, it allows/denies the access. PDP takes more information from PIP for decision-making, and PEP implements the decision made by PDP. The policies are maintained in the PAP.

Traditionally two access control mechanisms were used:

- Attribute-based access control model (ABAC)
- Role-based access control model (RBAC).

In RBAC, the third-party authority defines different roles for accessing the resources, and each role is associated with different levels of access rights. The role of a user is verified before giving access rights to the user. Based on their role, the user gets permission to use the resources; however, defining role and their policies makes this mechanism unsuitable for a dynamic environment [17].

In ABAC, access of resources is associated with the attributes of the user, and based on these attributes, access policies are defined. When user requests the access to the resources, user attributes are verified instead of their roles. Access is provided based on the type of attributes shared by the user. This mechanism is suitable for a dynamic environment as access is not depending on roles that are needed to predefine.

There are various variations made in access control mechanisms as per application requirements, and few of them are presented below and a comparison of these methods is presented in Table 2.1:

- **Bilayer Access Control (BLAC)**: The bilayer access control mechanism is a two-step method that is used to grant access requests: first, request is checked against subject rules and, in the second step, against object rules. BLAC uses the concept of pseudo-role (job function). Attributes are divided into two categories, static and dynamic, depending on access request attributes. The process of privilege and permission modification is complex [17].
- The identity capability-based access control (ICAC) [18] mechanism is based on the capability of the subject. Capability is calculated based on the identity of subject/device and access rights. To access the device, the requester device should provide its capability. Depending on the capability of device, access is granted/denied. This scheme provides scalable, time-efficient and secure access control but needs to apply an access revocation mechanism.
- Isolation-role-based access control (I-RBAC) [19] is the variant of role-based access control mechanism. Isolation role is added to a predefined set of roles. Whenever a delegate tries to access a resource, the first check is done on the primary roles; if unsuccessful, then the isolation role is checked, then access is granted if the role matched; otherwise, access is denied. This scheme is implemented for the health-care system. This mechanism provides two-step authorization.
- Context-aware trust role-based access control (CATRAC) combines context-aware access control and trust-based access control mechanisms to acquire benefits of both [20]. But the third-party involvement in defining roles increases the computational overhead.
- In health-care scenarios, task failure results in serious consequences. In such systems, the access right decisions are more dependent on the context of a current task than the role of the subject. While high priority to task role is also a vital parameter for decision-making, task and role access control mechanisms are merged in task- and role-based access control (TRBAC) mechanism [21]. But it is difficult to define roles and tasks.
- The task-attribute-based access control (TABAC) mechanism combines task and attribute access control methods to achieve benefits from both. This is dynamic in nature as uses attribute-based access control [22].

TABLE 2.1 Comparison of Access Control Mechanisms

ACCESS CONTROL MODEL	GRANULARITY	SCALABILITY	PARAMETERS		
			DELEGATION	COMPLEXITY	DYNAMIC
ABAC	Yes	Yes	No	Yes	Yes
RBAC	No	No	Yes	No	No
BLAC	Yes	No	No	Yes	Yes
ICAC	Yes	Yes	No	No	Yes
I-RBAC	No	No	Yes	No	No
CATRAC	No	No	Yes	Yes	No
TRBAC	No	No	Yes	No	No
TABAC	Yes	Yes	No	Yes	Yes

2.5 IMPLEMENTATION PERSPECTIVE

Authentication and authorization are mandatory steps before giving access to the resources. There are various ways to provide secure access to services; however, in the era of cloud computing, big data and IoT, it is important to have mechanisms that are decentralized and distributed. The centralized approaches may not be suitable for applications. The user provides attributes for authentication; however, he doesn't have any control over the disclosure of attributes as it is done through a third party. There is a need for architectures that can provide complete control of attributes in the hands of the user. The minimal disclosure of attributes is the solution for preserving the privacy of users' personal data. Self-sovereign identity (SSI) is the digital identity of the user, and he has complete control over it. In this, user can decide which attributes and how many attributes he should disclose to access the services. The user's birthdate, driving license or university degree can be used as SSI [23].

The OAuth-based authorization with the help of decentralized identities (DID) and verifiable credentials (VC) can be a suitable mechanism to provide complete control in the user's hands [23,24].

2.6 SUMMARY

Due to advancements in technologies, the increased use of online services has led to accumulation of a large amount of data. Many organizations are keeping their data on the cloud and providing access to these data through web services. On the other side, users can access these data from any place and at any time through any device. Most of the communication is done through wireless mode, and hence, it is prone to security attacks. Before communication, both end parties should know whether they are communicating to the legitimate partner. Hence, authentication plays an important role in the security of the services. Furthermore, authorization and access control are important to allow/deny access to the legitimate user based on the policies. In this chapter, an overview of authentication, authorization and access control mechanisms is presented. The difference between authentication, authorization and access control is also discussed in this chapter. The last part of the chapter discusses the various mechanisms of access control in detail. In this digital era, there is a need of a decentralized, privacy-aware access control mechanism. The various techniques like SSI, DID and VC are also discussed.

REFERENCES

1. N. Dey, G. Shinde, P. Mahalle and H. Olesen. (2019). *The Internet of Everything: Advances, Challenges and Applications*. Dè Gruyter, Germany.
2. G. Shinde and H. Olesen. (2015, April). Interaction between users and IoT clusters: Moving towards an Internet of People, Things and Services (IoPTS). In *Proceedings of WWRF Meeting 34*, Santa Clara, CA, Apr. 2015.
3. P. Samarati and S. C. de Vimercati. (2000, September). Access control: Policies, models, and mechanisms. In *International School on Foundations of Security Analysis and Design*, pp. 137–196, Springer, Berlin, Heidelberg.
4. V. C. Hu, D. Ferraiolo and D. R. Kuhn. (2006). *Assessment of Access Control Systems*. US Department of Commerce, National Institute of Standards and Technology, Gaithersburg, MD.
5. P. N. Railkar, P. N. Mahalle and G. R. Shinde. (2018, November). Access control schemes for machine to machine communication in IoT: Comparative analysis and discussion. In *2018 IEEE Global Conference on Wireless Computing and Networking (GCWCN)*, pp. 59–63, IEEE, India.
6. P. N. Railkar, P. N. Mahalle, G. R. Shinde and H. R. Bhapkar. (2019). 3. Threat analysis and attack modeling for machine-to-machine communication toward Internet of things. In *The Internet of Everything*, pp. 45–72, De Gruyter, Germany.
7. T. Hardjono. User-managed access (UMA) profile of OAuth 2.0. Available: http://docs.kantarainitiative.org/uma/draft-uma-core.html.
8. COM Authentication, at http://msdn.microsoft.com/en-us/library/aa913215.aspx.
9. http://openid.net.
10. W. A. Alrodhan and C. J. Mitchell. (2007) Addressing privacy issues in CardSpace. In *Third International Symposium on Information Assurance and Security*, Manchester.
11. OpenAM Project - About OpenAM. Available: http://openam.forgerock.org/
12. Specifications Overview | FIDO Alliance. Available: https://fidoalliance.org/specifications.
13. G. Alpár and B. Jacobs. Credential design in attribute-based identity management.
14. D. Baier, V. Bertocci, K. Brown, S. Densmore, E. Pace and M. Woloski. A guide to claims-based identity and access control. Microsoft Press.
15. *The Essential OAuth Primer: Understanding OAuth for Securing Cloud APIs*. (2013). Whitepaper, Ping Identity Corp., US.
16. Welcome to OpenID Connect. http://openid.net/connect
17. S. Alshehri and R. K. Raj. (2013). Secure access control for health information sharing systems. In *IEEE International Conference on Healthcare Informatics*, pp. 277–286, Washington, DC.
18. P. N. Mahalle, B. Anggorojati, N. R. Prasad and R. Prasad. (2012). Identity driven capability based access control (ICAC) scheme for the internet of things. In *IEEE International Conference on Advanced Networks and Telecommunications Systems (ANTS)*.

19. N. Gunti, W. Sun and M. Niamat. I-RBAC: Isolation enabled role-based access control. In *Ninth Annual International Conference on Privacy, Security and Trust*, Montreal.

20. C. Ghali, A. Chehab and A. Kayssi. (2013). Combining task- and role-based access control with multi-constraints for a medical workflow system. In *2010 10th IEEE International Conference on Computer and Information Technology*, pp. 1085–1089. West Yorkshire, IEEE.

21. I. J. G. Mallare and S. Pancho-Festin. (2013). Combining task- and role-based access control with multi-constraints for a medical workflow system. In *2013 International Conference on IT Convergence and Security (ICITCS)*, pp. 1–4. *IEEE*.

22. L. Yi, X. Ke and S. Junde. (2013). A task-attribute-based workflow access control model. In *IEEE International Conference on Green Computing and Communications and IEEE Internet of Things and IEEE Cyber, Physical and Social Computing*, Halifax.

23. P. N. Mahalle and G. R. Shinde. (2021). OAuth-based authorization and delegation in smart home for the elderly using decentralized identifiers and verifiable credentials. In *Security Issues and Privacy Threats in Smart Ubiquitous Computing* (pp. 95–109). Springer, Singapore.

24. P. N. Mahalle, G. Shinde and P. M. Shafi. (2020). Rethinking decentralised identifiers and verifiable credentials for the internet of things. In *Internet of Things, Smart Computing and Technology: A Roadmap Ahead* (pp. 361–374). Springer, Singapore.

Open Authorization 2.0

3

3.1 INTRODUCTION

OAuth 2.0 is an open standard for authorization. Considering OAuth at high level, it's not an application programming interface (API) or a service; basically, it is a standard that applications can use for secure access delegation. It is designed mainly for granting access to a set of resources, and it works over HTTPS and authorizes devices, servers, APIs, and applications along with access tokens instead of login credentials. OAuth 2.0 uses access tokens which have data that represents the authorization to access resources on behalf of the end user. Format of access token is not specified by OAuth 2.0, but in Widley JWT (JSON Web Token) is used. In Ref. [1], a detailed study of OAuth evolution and various techniques for implementation of web server application is provided. OAuth Core1.0 draft was released on October 3, 2007 by folks Blaine Cook, Chris Messina and their team; the idea of OAuth came into picture in the context that how Twitter API and OpenID connect can be used together for delegation and authentication. There are mainly three versions till date, namely, OAuth 1.0, OAuth 2.0 and the latest OAuth 2.1, which is recently consolidated from OAuth 2.0, but not yet released. The most commonly and practically used version is OAuth 2.0. If we see the specification among three, OAuth 1.0 differs a lot from OAuth 2.0 and also there is no backward compatibility provided. OAuth 2.1 carries similar specifications from OAuth 2.0 having some significant changes, which are required as per the current requirements.

OAuth 1.0 was released in April 2010 this was replaced by OAuth 2.0 in October 2012 having specifications RFC 6749 and RFC 6750. OAuth 2.1

DOI: 10.1201/9781003268482-3

has consolidated best practices implemented over the last eight years since OAuth 2.0 was released. OAuth 2.0 has numerous extensions due to addition of specifications in subsequent years. Most recently, on July 30, 2020, a new OAuth extension has been proposed by adding a specification. OAuth 2.1 dictates the best security practices and omits certain parts of OAuth 2.0, but the remaining OAuth specifications are retained.

Requirement of OAuth can be easily understood through examples. Let's consider an example that how OAuth can solve basic problem of granting limited access to third parties by giving user credentials. Nowadays, great comfort cars are launched with valet keys for the comfort of owners. These keys may be handed over to parking attendants, and but by using this key, no one can go more than a few kilometers as they are created with limited features. By using that key trunk opening also not possible, also to get access to address book feature will be denied. This limited feature with restrictions is imposed by that key to maintain vehicle safety and privacy considerations.

Gradually, a number of websites have been launched offering various services that bring more functionality together, including combination of multiple sites. Consider a website for facilitating prints of client's photographs according to need. In a research work [2] based on OAuth 2.0 protocol, the authors proposed a scheme for authentication for securing access to an Internet of Things (IoT) network. Many types of implementations can be possible to bring a solution to the end user by using address book of social networking sites [3]. Few APIs are used to build applications that are compatible with desktops. All these are really great services, but one important problem is that while implementing them, they require user credentials like password and username for another website. But in such cases, when you are ready to share your personal secrete credentials, you are exposing your passwords that you have used previously, for example, for online transactions; indirectly, it leads to giving whole access to websites that they can access your account any time and do anything they want. OAuth tries to solve this problem by giving limited access to your data, and also user's identity and credentials will not be disclosed. It permits the user for granting access of your resources, which are stored on one website to another website for some facilitation purpose. OAuth-enabled OpenID uses user's credentials to every website, but by use of token management system, eventually signing in is easily achieved in OAuth 2.0.

It is highly useful for software architects or developers for building:

- web applications
- browser-based apps
- desktop applications
- mobile applications.

OAuth is a way to get access to protected data from an application. It's highly safer and more secure as far as Login with password concept is concerned.

For API developers, basically it supports:

- mobile applications
- web applications
- server-side APIs.

OAuth allows to application developers to get access to user's related data without sharing their password very securely.

On the whole, OAuth 2.0 is an authorization framework or protocol for enabling third-party applications to get the limited access to any user accounts over HTTP services. This keeps user account credentials safe without divulging it to third parties. Mainly OAuth 2.0 focuses on client and developer's intelligibility along with providing specific flows for authorization to applications that support desktop, web, android and various remote devices. IETF OAuth Working Group plays a crucial role in developing specifications and its extensions.

3.1.1 OAuth Roles/Main Actors of OAuth2.0

OAuth defines mainly four roles:

- Resource owner:
    The end user who can authorize an application for accessing their personal account is resource owner. An entity is capable of granting access to a protected resource.
- Resource server:
    It is server that hosts the protected resources by using access tokens. It can also handle accepting and responding to protected resource requests.
- Client:
    It is an application that accesses users' account. Prior to it, it has to be authorized by the respective user, and it is mandatory that API should validate the authorization process.
- Authorization server:
    Access token is issued to clients by the authorization server, main role of this server, only after authentications with resource owner; authorization will be gained after it. This server plays an important role in the authorization process.

According to OAuth specifications, it is not only resource server communicates to authorization server with some constraints. Here both resource and authorization servers could be separate or same entities. Here both resource and authorization servers could be separate or same entities. Access tokens are issued by single authorization server and are accepted by multiple resource servers.

3.2 MOTIVATION

Before OAuth, traditional client server authentication model was in existence, and it had to face several problems and limitations.

- Resource owner's credential was stored by third-party applications for future use. Mainly password was stored in clear text. So keeping password at third part apps leads to breakage of security policies. Access was given to protected resources by third-party applications. Therefore, it was an open invitation for hackers to steal confidential information.
- Access to third parties could not be revoked individually; rather, it was required for resource owners to track all the third parties and cancel at single time. This was achieved through changing password directly.
- Password authentication was required by servers though the security weaknesses pristine in passwords.
- Data protection was solely dependent on the end user password and also whole protection was dependent on it, so compromising third-party application leads to compromising end user's password.

All abovementioned issues are taken into consideration and OAuth introduced one layer for authorization. Also it has separated important roles of resource owner from that of client. Client can request for accessing resources by issuing different types of credentials, resource server hosts resources that are controlled and managed by resource owner. In OAuth, access token plays an important role for granting permissions and knowing identity over network [4]. Access token has access attributes, specific scope, and also lifetime of it. Eventually to access protected resources without using credentials that is present with resource owner. In Ref. [5], the author introduced advanced methods for token binding mechanism for enhancing the security of APIs. Access token is gained by client. The third-party clients get an access token provided by the authorization server by the approval of resource owner. Client uses the access token for accessing some protected resources that could be hosted through resource server. In a research work [6], structured analysis of user access privacy properties is mentioned; also it has described simplified way to identity provider for tracking user accesses and also proposed different methods for mitigating privacy issues to some extent. In Ref. [7], for sharing web resources, capability-based access control mechanism has been provided, which is based on OAuth 2.0.

For example, consider a banking application that is used to help you in making transactions, withdrawals, credits and debits and managing your budgets and that lets you access a number of bank accounts. Suppose you want to manage four bank accounts of yours and that too with the application; you have different usernames and passwords for each. Now for accessing bank account through the application, it requires your sensitive credentials like username, login password, transaction password, etc. and you are giving all sensitive information to the app. In such case, how to rely on an app that can be hacked by a hacker. So to overcome this, you have to change the password frequently for each account, but this could not be a feasible solution. This fundamental problem is addressed by OAuth 2.0.

OAuth 2.0 facilitated resource-sharing information very securely with the help of secure authorization token exchange mechanism. For third-party application, access delegation and resource sharing over network, OAuth has developed a standard protocol to figure out such problems.

3.3 PROTOCOL OVERVIEW

Protocol works around mainly four actors of OAuth2.0 and how they communicate to each other. By using OAuth 2.0, access requests are first initiated by the clients for, e.g., website, desktop application, mobile application or any smart TV apps. The exchange, request and response of tokens have standard flow explained below:

- Authorization is requested by client from resource owner. Resource owner may be granted authorization request, or this may be performed by authorization server.
- An authorization grant is received by client, extension grant which is one of the main grants out of four grant types used here for representation of resource owner authorization as a credential. The authorization grant type relies on client's method for requesting

validation and the types which are compatible by authorization server for authorization.

- Authorization grant is presented after the client request for access token; it is carried out by authenticating and communicating to authorization server.
- Authentication of client is done by authorization server, and also the authorization grant's validation tasks are taken care by authorization server after validation token gets issued to the client.
- Authentication of client and resource server is achieved through access token, which is further utilized by client for getting access of protected resources from resource server.
- Eventually, access token validation is done through resource server, and it serves the request if the token is valid.

The detailed communication flow between client, authorization server, resource server and resource owner is illustrated in Figure 3.1.

In an abstract protocol flow [7], we can see one side client and other side resource owner, authorization server and resource server. Communication between client and both the authentication server and resource server along with resource owner is obtained. The main purpose of this flow is delegating authorization over network. In a research work [8], access delegation and authorization challenges in IoT-enabled smart house for aged people with the presence of constrained devices are discussed. By considering a guest person as a visitor, use case is discussed. For delegating access, verifiable credentials and decentralized identifiers are used.

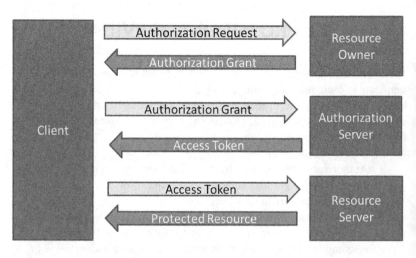

FIGURE 3.1   Abstract protocol flow

3.4 USE CASE

We shall consider two use cases for OAuth implementation: user agent and web server. In user agent [9] use case, implementation of OAuth 2.0 is illustrated with educational application along with additional server and authorization server.

3.4.1 User Agent as Use Case

3.4.1.1 Educational Application

Sulochana has installed an educational application on her desktop machine. She maintains the database of students she teaches. The database contains student records, results, attendance, etc. She has to upload the same on one of the official website of government, named www.zpschool.test.com.

So for uploading Sulochana's student database on the official site, her installed educational application obtains the access to database by using the authorization mechanism.

Conditions Required Before
- Application executed in JavaScript language has been installed by Sulochana in her computer, which should run properly in her browser. The website www.zpschool.test.com gets an access token by using OAuth technique.
- There should not be the case that application be supported by any website. Therefore, installed application requires updating database by its own.
- It is required to register installed application with www.zpschool.test.com.
- For authentication and identification purposes, Sulochana has registered on educational school website www.zpschool.test.com.
- Sulochana's web browser shall be able to communicate with additional web browser named www.sopan.test.com. Additional server has ability to provide extraction of access token in the fragment of URL, which is from the script.

Post-conditions
- Access token is received by Sulochana's web browser after successful implementation.
- www.sopan.test.com as an additional server redirects URL's fragment; www.zpschool.test.com sends the access token. Here

Sulochana's web browser should obey redirection by keeping fragment in it. Sulochana's web browser starts downloading scripts, and access token gets extracted from the fragment part. This is from additional web server present at www.sopan.test.com. The access token made available for educational application by using access token application can be used to obtain access to Sulochana's data present at www.zpschool.test.com.

Mandatory Artifact
- The website www.zpschool.test.com possesses an application, which should be registered with the application, which runs in Sulochana's web browser for only identification purpose.
- Application that is running at www.zpschool.test.com should give response to access token after Sulochana's authorization.
- It is mandatory to have Sulochana's authentication with www.zpschool.test.com.

3.4.2 Web Server in Web Application

In Ref. [10], use case of photo printing service application along with third-party application is explained.

Illustration – Satej creates an account in a web application at www.saiphotos.test for ordering and printing his photographs that are saved in www.photostore.test server. Web app at www.saiphotos.test then receives Satej's consent to access his photos without any credentials that are required for authentication at site www.photostore.test.

Pre-state
- The first condition is that Satej has already registered with www.photostore.test for activating the authentication.
- There is web application present on www.saiphotos.test that has set up authentication with another application at the domain www.photostore.test that is required for further resource utilization.

Post-state

After the valid authentication, authentication code will be sent by www.photostore.test to the www.saiphotos.test. In order to get access token, authorization code will be used by the app hosted by www.saiphotos.test, from www.photostore.test. After validating authorization code that has been already submitted and authenticating that app present at www.saiphotos.test, access token is issued by the app present at www.photostore.test. Eventually, access token will be used by app present at www.saiphotos.test, for obtaining access to Satej's images that are stored at www.photostore.test.

For short-term access to application, access tokens get expired after its role in flow. So services at www.saiphotos.test required to replicate similar procedure of OAuth for obtaining Satej's authorization, for accessing his photos present at app at www.photostore.test. For long-term access delegation, when Satej feels to do so, by giving access to his own resources stored at www.photostore.test, long-lasting tokens can be used and issued by authorization server that deals with www.photostore.test.

Basic Requirements
- www.photostore.test, the application present has to authenticate Satej. The method of authentication may be out of scope of OAuth specifications.
- It is mandatory to issue HTTP redirect requests in browser application of Satej. This task is carried out by OAuth client by the main server named www.saiphotos.test.
- The server www.saiphotos.test, which hosts an OAuth client, must be capable of issuing the HTTP redirect requests to Satej's browser.
- Satej's unauthorization for further access to his photos by www.saiphotos.test has to be gained by the www.photostore.test application.
- Satej doesn't have to manually involve in the process of authorization by OAuth, for example, entering password and typing any URL.
- Authentication of application at www.saiphotos.test with app present at www.photostore.test shall be carried out. Prior issuance of access token validation shall be carried out for authorization code.
- Every time the application at www.photostore.test might recognise to Satej, scope of accessing the site www.saiphotos.test has requested while asking for Satej's authentication.
- Sometimes Satej's authentication with the www.photostore.test may not be in the OAuth scope. It is prerequisite for Satej to register at www.photostore.test.

3.5 AUTHORIZATION PROCESS

Authorization plays a crucial role in the OAuth framework. It basically deals with different types of grants. In recent draft from ietf.org, which is released on 5 October 2021, the specification is introduced with new modification over old version of OAuth 2.0 from RFC-6749 [11]. Authorization process mainly includes four grants types:

- authorization code grant
- implicit grant
- resource owner password credential grant
- client credentials grant.

The authorization grant is used by client for requesting access token. Extension mechanism is also provided for defining additional grant types. OAuth2.0 draft introduces four grant types that are implicit grant, authorization code grant, client credential grant and resource owner password credential grant. Also, it provides a method for extended functionality, required for supplementary categories. To understand OAuth 2.0 authorization process, it is required to know authorization code grants and its types.

3.5.1 Authorization Code Grant

The first grant in the authorization process is authorization grant; it is a credential representing authorization to access protected resources which are used by the client to obtain an access token. The access token and refresh token are obtained through the authorization code grant, and also it is improved for the confidential clients. As this flow is redirection based, client must have capability to interact with resource owner's user agent, that is web browser, and client must be capable of receiving incoming request from the authorization server.

3.5.1.1 Authorization Code

It is an intermediary between resource server and client; it can be gained through authorization server. This code basically provides few important security benefits like resource owners, and user agent is bypassed in the scenario of access token transmission; it is directly transferred to the client. In Ref. [12], four extended authorization grant flows for issuing an interoperable access token (IAT) are presented. Under different types of domain, IAT has global scope for accessing it. Interoperability could support different domains with the authorization as a service. Apart from original grant flow, it has not only local scope but also global access. Instead of requesting authorization directly from the resource owner, the client can direct the resource owner to an authorization server. This directs the resource owner back to the client with the authorization code. By using authorization code, authentication is carried out by authentication server with the resource owner; this obtains authorization before directing the resource owner back to the client with the authorization code. This happens because in authentication process authorization, server has to authenticate resource owner. Client is unaware of resource owner's credentials.

This flow involves the following steps as depicted in Figure 3.2:

1. Initially client starts this flow; user agent of resource owner will be directed to the authorization end point of authorization server. Some local states, client identifier and requested scopes are mentioned by the client. It also includes redirection URI, where the authorization server sends back user agent after the access status, whether it is granted or denied.
2. Resource server is authenticated by authorization server through the user agent. It sets up depending on grant or deny status of client access request which is initiated by resource owner.
3. Authorization server redirects the user agent back to the same client by using the redirect URI which is issued recently while registration of client by assuming resource owner-granted access. The redirection code is included with authorization code and local state provided by the client.

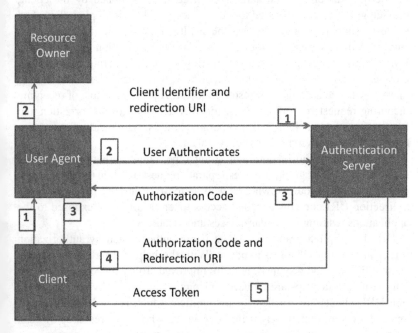

FIGURE 3.2 Authorization code flow

4. From the authorization server's token endpoint, request of client is placed with the help of authorization code accepted previously. At the time of making request, authentication process needs to be carried out and client has to authenticates with the authorization server. Redirection URI is included by client, which is used for getting the authorization code in verification process.

5. Authentication of client by authorization server validates authorization code, and it makes sure that the redirected URI received matches with the URI used for redirecting the client in step 3, and if it is valid, then the authorization server will respond it back with an access token and optionally a refresh token.

Steps 1–5 cover the entire authorization code flow.

3.5.2 Implicit Grant

This grant simplifies authorization process that optimizes a client who has been implemented in browser with the help of scripting language, like JavaScript. Rather than an authorization code to be issued by the client, implicit grant type can be used for getting access token, in implicit flow. And it doesn't support issuing of access tokens. It is functionalized for public client; basically, it's identified for operating specific redirection URI over web. Java scripting language is used for creating and implementing client in web browsers. Due to redirection-dependent flow, here client should be able to interact with the user agent of resource server. Also it is capable of receiving incoming request from the authorization server. Here for authorization and for an access token, client makes separate requests. As a result of the authorization request, client receives access token.

Unlike the authorization code grant type, client authentication is not included in implicit grant, it makes separate request and it is totally depends on existence of resource owner as well as redirection URI registration. As redirection URI contains encoded access token, it might be exposed to the applications residing in similar devices and resources.

In this implicit grant flow, client is not authenticated by authorization server at the time of token issuance. With the help of this implicit grant, client efficiency and responsiveness are improved. To get an access token, the efforts of circular trips are reduced. In some of the cases, by using redirection URI, clients are verified for recognition, which is used for delivering an access token to client. By using user agents who are of resource owners, access tokens are visible to other applications. Nonetheless, this convenience must be measured against the security insinuations, after the use of implicit grants, and mainly if the authorization code grant type is available.

Figure 3.3 illustrates implicit grant flow as follows:

1. Implicit grant flow gets initiated by client, and user agent of resource owner gets directed to the authorization server's end point. Some local states, client identifier and requested scopes are mentioned by client. It also includes redirection URI, where the authorization server sends back user agent after the access status, whether it is granted or denied.
2. This step focuses on authenticating resource owner by authorization server. It sets up communication depending on the consent of resource owner as granting or denying access request from client.
3. Considering that resource owner has granted access and user agent is redirected back at client, this is initiated by authorization server. By using redirection URI which is given previously, Access token is comprised in the same URI.

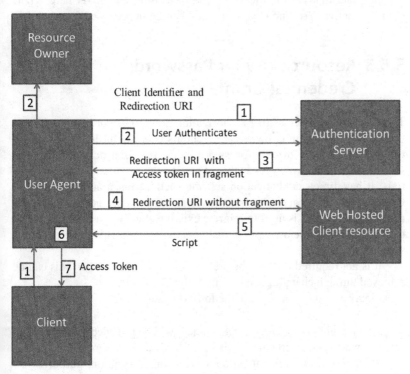

FIGURE 3.3 Implicit grant flow

4. Request is placed to web-hosted client resource by the user followed by redirection instructions.
5. Web page gets returned by web-hosted client resource, which is able to access the entire redirection URI along with fragment, kept by the last user agent. Also, it draws out parameter along with access token included in the same fragment.
6. In this phase, withdrawing of access token is carried out. This is achieved locally by web-hosted client resource; this provides script for execution purpose by the end user.
7. Lastly, clients get access token that is sent by the user agent.

Steps 1–7 represent the entire flow of implicit grant.

In Ref. [13], the authors proposed the MyDataChain framework for improving current specification of OAuth by using block chain technology, as there is interaction gap between resource server and authorization server. This proposed framework could gain the expected results by it, by not holding any data that can be identified from the Blockchain. As here NIZK (non-interactive zero-knowledge) method is used in proposed framework.

3.5.3 Resource Owner Password Credential Grant

To get an access token as like authorization grant, password and username in the form of resource, owner's credentials could be utilized. By use of this grant, resource owner's password credentials can be obtained, it is suitable for clients which is capable of getting the resource owners credentials. It has direct authentication scheme such as digest authentication to OAuth; this can be achieved by converting the stored credentials into an access token, which is used to migrate existing clients. If a high-level trust between the resource owner and the client exists, then only the credentials should be used.

It is not required to store the credentials of resource owner; this can be achieved through changing credentials with refresh token.

Figure 3.4 shows the flow in the following steps:

- For login process requirement, client will be provided by resource owner with its credentials like password and username.
- In client and authentication server communication, authorization server provides the request token requested by the client; it includes whatever credentials that are received from the resource

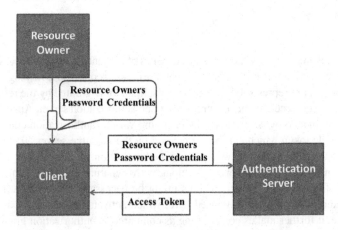

FIGURE 3.4 Resource owner password credential grant flow

owner. Client has to authenticate with the authorization sever while placing the request.

- Eventually, it's time to get an access token, client gets authenticated through authorization server and it validates the credentials of resource owner, and if it is valid, then access token will be issued.

3.5.4 Client Credentials Grant

Authorization scope is limited for the protected resources that are controlled by client. For an authorization grant, credentials of client need to be used.

Only when there are confidential clients, they have to use this grant type for authentication purpose. In this flow, basically credentials of client are used as an authorization grant.

Figure 3.5 shows the flow with the following steps:

- Access token from the token end point is requested by the client after authenticating it with the authorization server.
- Authentication of client is approved by authorization server and if client is authentic, then issued an access token to it.

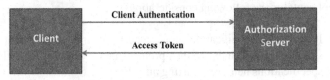

FIGURE 3.5 Client credential grant flow

3.5.4.1 Types of Token

3.5.4.1.1 Access Token

Access token is string, which means it consists of credentials used to access protected resources. It is not visible to client. Particular scope and how long it will be accessed are represented by the token itself, and it is granted by the resource owner, and enforced by the resource server and authorization server. Abstraction layer is provided by access token by changing various authorization constructs with single token which can be understood by the resource server. Tokens are nothing but identifiers for withdrawing authentication. Tokens might contain whole information of authorization along with signature and data. Across the scope of its specification, credentials might be necessary for the client to use the token. The abstraction is used here, which activates access token issuing scheme, it further makes constraining as compared to authorization grant, this can be is used to get it. It also clears the need for resource server to understand a wide range of various authentication methods. Depending on the security need for resource server, various formats of access tokens can be used. Sometimes, cryptographic properties as a utilization method are taken into consideration.

3.5.4.1.2 Refresh Token

Refresh token credentials are used to obtain access tokens. Authorization server holds the right of issuance of refresh token. Authorization server issues an access token to the clients, and it is used for getting new access tokens whenever the current access token expires or becomes invalid. Refresh token is included only at the time of issuing access token; here, refresh token will be issued by authorization server. It is a string that represents authorization granted to the client by the resource owner. Similar to access token string, it is not visible to the client. To retrieve the authorization, information token denotes an identifier.

3.6 SECURITY ANALYSIS

OAuth 2.0 protocol specifications given in RFC 6749 contain various security concerns based on real-time scenarios. Generic threat model of OAuth 2.0 protocol contains the following considerations:

- It represents specifications concerned to security aspects that are suitable to implement OAuth protocol and eventually how these specifications help in thwarting any attacks.

- Scope and documents and any assumptions are considered while creating the security model.
- It provides the best suitable model for OAuth. This is used to describe related corrective measures to fight against threats and attacks.

Threats are nothing but attacks on tokens used in OAuth and resources that are protected by tokens of OAuth in the flow. There is high risk if security concerns are not implemented in the respective places of the OAuth model. Each part of the protocol is developed securely by considering real-time scenarios of threats, for example, structuring of threats along with the protocol structure. The following are well-known considerations that must be considered while implementing authorization server. In Ref. [14], the authors tried to implement one proposed method, which is based on email authentication, and it reduced the success rate of authentication by hackers to less than 0.8%. Different kinds of attacks are taken into considerations to find whether OAuth 2.0 is susceptible to such types of attacks. Attacker assumptions and architectural assumptions need to be considered on a scenario basis; let's consider phishing attacks over OAuth server and how this can be handled.

3.6.1 Phishing Attacks

In these attacks, the attacker creates a website that is identical to the site which the victim uses for their services and has fields identical to the attacker's website, such as fields of username and password. In this case, phishing attack is a potential attack on the authorization server. The attacker can trick the user to visit the web pages by various means. It can only be identified by verifying the web address in the address bar; otherwise, the user would enter their credentials like username and password in respective fields of this website as its contents are similar to those of the original website. The user is tricked by the attacker using fake Barmecidal server. The web page is embedded with the authentic web page which is then phished as a new look in real application. This mixed web page doesn't show the address bar. Therefore, there is no option remains with user for confirming whether the page is original one with application. Unfortunately, this happens in mobile applications more frequently. Many times, OAuth providers encourage third-party applications for providing local application, rather than embedding authorization page allowance, to see that in web scenarios.

3.6.2 Countermeasures

To overcome such type of phishing attacks over web pages, it is required that authorization server should be served through the HTTPS for avoiding DNS spoofing. Risks of phishing attacks could be avoided by educating developers who work for building an authorization server. Various steps could be taken for preventing web pages from embedding it with native applications. In Ref. [15], the authors suggested a novel approach called VOAuth for defeating phishing attacks. VOAuth introduces a validation mechanism in OAuth authorization. This enhances OAuth to process faster. A combination of validation client and validation gateway in the validation system brings the assurance of authentic authorization server. Therefore, password storage scenario will not be raised for long term, which eventually leads to protection against phishing attacks. Eventually, VOAuth model has been implemented successfully and used for Internet applications for mobile devices. After the user's experience, no fishing attack is reported. This shows the successfulness of VOAuth.

3.6.3 Clickjacking

One more type of attack that is frequently occurring is clickjacking; here, the user is tricked by the attacker to click on a window that popes up into web pages that are totally disguised sometimes or may be visible. Due to this, the user may be tricked in different ways to download any kind of malware or get redirected to malicious web pages, which may lead to, for example, leakage of credentials and loss of money. In Refs. [16,17], clickjacking attack is discussed and addressed. Authorization request is an important factor, as it get susceptible to clickjacking attack, But authorization server handles it and plays an important role in clickjacking; it must protect users data by preventing clickjacking, many countermeasures are provided in the OAuth specification. Use of frame busting JavaScript, response header of HTTP, X-frames option in the implementations are provided in OAuth specification as a countermeasures. Importantly, CSP (content security policies) above level 2 shall be used at authorization end point to avoid such clickjacking attacks to user. In case, remaining authorization end point should be used to client authorization and authenticating the user. As a user agent, CSP are supported to it, for preventing from malicious domains. To make required changes in configuration by admins should be permitted by authorization servers for better security. This way clickjacking could be avoided by using the above techniques in authorization flow.

3.7 SUMMARY

To summarise OAuth 2.0 framework, when backward compatibility point is concerned, it doesn't support the OAuth 1.0 version. It changes signatures with HTTP for all the communications. OAuth 2.0 has a number of variances in implementations; for integrating with the vendor, sometimes custom code is required. In the documentation and specification part, it is said that OAuth 2.0 is not a protocol, but rather, it is a framework. For access delegation to API, authorization framework OAuth2.0 is used.

Mainly it includes the client who requests a particular scope that gets authorized by resource owner or it gives permission to do it. Basically as per the flow dependency, authorization grants are exchanged for refresh tokens and access tokens, and various scenarios for authorization and client variances can be addressed in OAuth 2.0.

REFERENCES

1. M. Darwish and A. Ouda. (2015, October). Evaluation of an OAuth 2.0 protocol implementation for web server applications. In *2015 International Conference and Workshop on Computing and Communication (IEMCON)*, pp. 1–4, IEEE, Vancouver, BC, Canada.
2. J. Khan, J. Ping Li, I. Ali, S. Parveen, G. Ahmad Khan, M. Khalil and Shahid, M. (2018, December). An authentication technique based on OAuth 2.0 protocol for internet of things (IoT) network. In *2018 15th International Computer Conference on Wavelet Active Media Technology and Information Processing (ICCWAMTIP)*, pp. 160–165, IEEE, Chengdu, China.
3. P. Hu, R. Yang, Y. Li and W. C. Lau. (2014, October). Application impersonation: problems of OAuth and API design in online social networks. In *Proceedings of the Second Association for Computing Machinery Conference on Online Social Networks*, pp. 271–278, New York, NY, United States.
4. W. Li and C. J. Mitchell. (2020, September). User access privacy in OAuth 2.0 and OpenID connect. In *2020 IEEE European Symposium on Security and Privacy Workshops (EuroS&PW)*, pp. 664–6732, IEEE, Genoa, Italy.
5. Siriwardena, P. (2020). OAuth 2.0 token binding. In *Advanced API Security*, pp. 243–255, Apress, Berkeley, CA.
6. Yang, F., & Manoharan, S. (2013, August). A security analysis of the OAuth protocol. In 2013 IEEE Pacific Rim Conference on Communications, Computers and Signal Processing (PACRIM) (pp. 271–276). IEEE. Victoria, BC, Canada.
7. Abstract protocol flow(APF) of OAuth 2.0, at https://datatracker.ietf.org.

8. P. N. Mahalle and G. R. Shinde. (2021). OAuth-based authorization and delegation in smart home for the elderly using decentralized identifiers and verifiable credentials. In *Security Issues and Privacy Threats in Smart Ubiquitous Computing* (pp. 95–109). Springer, Singapore.

9. User Agent as a USE CASE, at https://datatracker.ietf.org/doc/html/draft-ietf-oauth-use-cases-01#page-6.

10. Web Server as a USE CASE, at https://datatracker.ietf.org/doc/html/draft-ietf-oauth-use-cases-01#page-3.

11. T. Lodderstedt. (2021). The OAuth 2.1 authorization framework draft-ietf-oauth-v2-1-04. https://datatracker.ietf.org/doc/draft-ietf-oauth-v2-1/

12. S. R. Oh and Y. G. Kim. (2020). AFaaS: Authorization framework as a service for Internet of Things based on interoperable OAuth. *International Journal of Distributed Sensor Networks*, 16(2), 1550147720906388.

13. S. -C. Cha, C. -L. Chang, Y. Xiang, Z. -J. Huang and K. -H. Yeh, "Enhancing OAuth with Blockchain Technologies for Data Portability," in IEEE Transactions on Cloud Computing, doi: 10.1109/TCC.2021.3094846.

14. C. J. Chae, K. B. Kim and H. J. Cho. (2019). A study on secure user authentication and authorization in OAuth protocol. *Cluster Computing*, 22(1), 1991–1999. Springer.

15. M. Xie, W. Huang, L. Yang and Y. Yang. (2016). VOAuth: A solution to protect OAuth against phishing. *Computers in Industry*, 82, 151–159.

16. Sancho, J., García, J., & Alesanco, Á. (2021). Authorizing Third-Party Applications Served through Messaging Platforms. Sensors, 21(17), 5716.

17. Shuai, L. I. A survey on security analysis of OAuth 2.0 framework.

User-Managed Access

4

4.1 INTRODUCTION

User-managed access (UMA) is a standard authorization protocol that has been released by Kantara Initiative. Kantara Group works over globe for promoting business relations based on data privacy and digital identity management [1]. UMA is one step ahead of OAuth 2.0, as a distributed set of resource servers can be controlled over an authorization server. The basic purpose of UMA is that it is a mechanism to control access to application programming interface (API). Hence, the role of UMA starts when API is called by any client. Many times after the post authentication API is called, user that might be authenticated with Open ID connect. For example, consider a mobile application; when you start it, it triggers and API is being called. So here the client is mobile application, and the app calls API. In UMA there is no token present at first.

UMA is actually built right on top of OAuth, while in its progressive journey, UMA contributors collected lots of issues and experiences. After all, eventually UMA 2.0 was released with a modified specification draft in February 2018. At the time of release, it was supposed to be technically stable, as the throughout the journey, it had to move with a lot of ratification process. Its journey began a few years back in 2015 for writing draft for UMA, but in March 2015, UMA V.1.0 ratified as recommendation, and later in December 2015, V.1.0.1 was published [2]. Approving access request from requesting party, with presence of resource owner, gets removed here in new policy. Depending on the policy defined in resource server, the authorization server can take the decision for access request handling mechanism.

In January 2018, the final recommendation was published by Kantara Initiatives and the logo of UMA 2.0 was designed by Domcat. In this draft, UMA business model report was published. Basically, this facilitates the use of data and access which is being shared by various third parties. In Ref. [3], the authors gave a more detailed idea about working of UMA 2.0 authorization. Asynchronous authorization can be handled in UMA, an important characteristic of UMA. To explain this, let's consider one example based on asynchronous authorization. Ramesh may request access for services to Ganesh, but Ganesh may not approve his request until he is online. So, at the authorization server, Ganesh could form a policy that can facilitate Ramesh to access the data anytime. There is no need of Ganesh to be online always for granting access.

UMA 2.0 protocol consists of the following specifications:

- **Federated Authorization for UMA 2.0**: This optional specification can be used with an UMA grant. Resource server is federated, and it defines a means for an UMA-activated authorization server in the context of authorized resource owner.
- **UMA 2.0 Grant for OAuth 2.0 Authorization**: When it is time to authorize access by resource owner, this grant can be specified by knowing how client could use a permission ticket for requesting an OAuth 2.0 access token for obtaining the access to a protected resource.

These specifications are used to set up access management and delegation for UMA 2.0 features and who wants to develop the following types of application:

- desktop applications
- web applications
- browser-based apps
- mobile applications.

Let's consider a banking scenario, where Satej as a bank customer is having an account in XYZ bank and Satej could share his account information with his brother Sai with limited access. Here, Satej is the resource owner and bank account service is the resource server; here, sharing management service is nothing but authorization server authenticates Sai for accessing protected resources from bank server. Bank workers have different client applications for their account management system. Satej's bank account is protected resource, but scopes are different for viewing account information and another is for performing transaction operations. What is special about UMA 2.0 when compared with OAuth 2.0 is it is a protocol built on top of OAuth 2.0. OAuth 2.0 protocol is extended by UMA 2.0 specifications and delegation of authorization server grant with consent to requesting party on behalf of resource owner for authorizing who can get access to their data and how long the access can be retained.

4.1.1 Roles of UMA Protocol

UMA has mainly five components (which are presented in the following subsections with respective example); each component plays a specific role for the entire execution of protocol.

4.1.1.1 Resource Owner

It as an entity whose user accounts in an application and it owns resources stored on that application. For example, Anvit has his account in video files storage application (in memory layer). In some cases, resource owner can be a nonhuman entity which is treated as a person for limited legal purposes.

4.1.1.2 Client Application

It is a web or mobile application that can access your resources on your behalf. For example, Anvit uses a video editing application called filmmaker "F" to access and edit videos stored in memory layer. Here, "F" is the client application.

4.1.1.3 Authorization Server

It authenticates resource owners using known credentials; examples of popular authorization servers are Facebook and Google. It also protects resources hosted at resource server.

4.1.1.4 Resource Server

Resource server stores the actual resources present and also makes it available for use. The memory-linked applications also act as resource server. For example, Google Drive, Facebook and Dropbox are some popular recourse servers present in the market.

4.1.1.5 Requesting Party

Requesting party is legal party or legal party, using client an application to seek access to protected resources. The requesting party may or may not be the same as the resource owner. For example, Anvit has shared his video collection to Satej; after that, Satej will act as a requesting party when he will attempt to access those videos.

4.2 MOTIVATION

Primarily access delegation is highly in need for developing web-based applications as resource owner requirements are considered. By the consent of resource owner, it can allow client or third-party applications to access protected resources after authorization process. When it comes to access delegation to user or client in web application, it is important to have trust certificates to know that user is not malicious. Implementers use various technologies to develop web apps that possess access delegation mechanism. But end users' (resource owners') data security considerations cannot be predicted. Based on the need in application, delegation could be of two types: user or administrative delegation.

Taking this problem into consideration, many recent techniques that solve security constraints have come to market. User may register application according to his/her needs, but when it comes to access delegation to third-party applications, user has to rely on privacy mechanisms applied to respective applications. Now in many cases, the user may give consent to third-party apps by trusting on them, but later there are chances that their secrete information get stolen by such apps. OAuth is one of the protocols that took such problems into considerations and started working on it with a large group. It released the latest draft of protocol in the market named OAuth 2.0. In this, the user delegates access to their resources as part of application they use. Unlike OAuth 2.0, UMA protocol has focused both technologically and conceptually on the problem of access delegation. In UMA 2.0, a user is permitted to delegate access to part of application that somebody else can use it with access permissions with the consent of resource owner.

This mechanism of delegating access obtains high chances of security, until resource owner doesn't allow the access to client as a third- party application till then access cannot be given to third- party application, indirectly to the client.

4.3 PROTOCOL OVERVIEW

UMA protocol flow differs from OAuth 2.0 flow as OAuth 2.0 has four roles and UMA 2.0 introduces a new role, i.e. requesting party. UMA protocol basically comprises two specifications: first is UMA grant, which is all about communication between authorization server and client. Client can get valid access token that client itself can use it for further accessing the protected resources from resource server. So we can say that UMA grant is basically the mandatory specification in UMA protocol.

The second specification is UMA-federated authorization recommendation; this specification communicates between authorization server and resource server. Resource server keeps the resources in the authorization server and it analyses the tokens provided by the clients.

Mainly UMA 2.0 defines five actions, namely, manage, control, protect, authorize and access. Kantara Initiative Working Group plays a crucial role in the development of specifications and its extensions from OAuth 2.0. In Figure 4.1, we can see that the resource owner has to deal with two actions, which are control and manage, concerning to authorization server and resource server, respectively. Manage action manages resources that are on the resource server. Control action deals with resource owner, where it controls the accesses given to registered resources. It creates some policy decision making on the authorization server. Here, granting consent by the resource owner will be in asynchronous manner unlike resource request time. Therefore, with the help of requesting party token (RPT), requesting party can access the data. Next, protect action deals with authorization server and resource server which are not tightly coupled in the architecture of UMA. Protect action protects the API

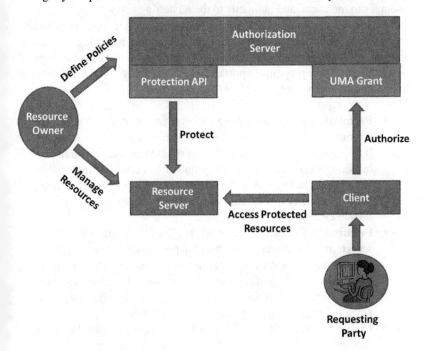

FIGURE 4.1   UMA components and flow

end points by means of PAT (protection API access token). This is a token with some specific scope of UMA protection by OAuth 2.0.

Protection API token is on the authorization server side, but it protects both the end points and eventually establishes relationship between resource server and authorization server. Authorize action deals with client and authorization server; this is the main action that deals with UMA grant on authorization server as it plays an important role in authorization between requesting party and authorization server. It uses RPT that is unique for all actors in the frame, which are resource owner resource server, authorization server, client and requesting party. When it comes to interaction between resource owner and requesting party, they use the application whenever they want. Therefore authorization delegation and party-to-party data sharing enabled by interaction. Authorization server allows granting consent asynchronously as authorization server policy decides granting consent which is required to resource owner to grant it. Lastly, access action deals with client and resource server. RPT is used here in communication, and this token is presented by the client. Resource server checks the validity of RPT. This is basically a claims token having permissions that are time limited. If the RPT is valid and if, in case, it has sufficient permissions, then the resource server is ready to return protected resources to the client, and indirectly to the requesting party.

Figure 4.2 illustrates the flow of grants, messaging paths, artifacts and communication between UMA actors. In this, claim gathering and claim pushing are represented, but at a time, only one can be used in one chain of events. Every entity plays an important role in the UMA. The following are some impressions as the flow of UMA is concerned:

- **Permission**: It is about giving access to resource. Permissions are strongly hidden to the clients and are part of authorization server. It takes active participation in authorization process through permission ticket. Resources are bounded to certain policies; they have scopes defined in it. So when it is time to authorize access to such resources, they are bounded with scopes with particular resources only.
- **Permission Ticket**: The code written inside the string that represents request permission is nothing but permission ticket. Permission ticket is primarily created by authorization server, but in the communication flow, client represents it, that is received from resource server. Finally, it reaches the token end point through the client, when the requesting party initiates to get redirected.
- **Claim**: To collect claim and access it is part of authorization process. Where claims are statements, these may have a number of attributes of different entities. In order to protect resources, claim collection

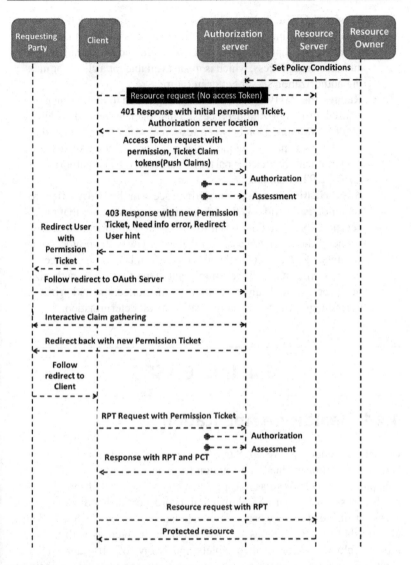

Set Policy Conditions

Resource request (No access Token)

401 Response with initial permission Ticket,
Authorization server location

Access Token request with
permission, Ticket Claim
tokens(Push Claims)

Authorization

Assessment

403 Response with new Permission
Ticket, Need info error, Redirect
User hint

Redirect User
with
Permission
Ticket

Follow redirect to OAuth Server

Interactive Claim gathering

Redirect back with new Permission Ticket

Follow
redirect to
Client

RPT Request with Permission Ticket

Authorization

Assessment

Response with RPT and PCT

Resource request with RPT

Protected resource

FIGURE 4.2 Example flow of UMA 2.0

(client and requesting party) based on the assessment leads to the authorization. Claim collection can be carried out in two ways in UMA: first by claim pushing and second by iterative claim gathering.

- **Claim Token**: String containing number of parameters for identification, collectively called package of claims. This will be received by authorization sever through the client in claim pushing process.

- **Persisted Claim Token**: This token made available itself for process flows in pipeline. A correlation handle issued by an authorization server that represents a set of claims collected during one authorization process, which is made available for a client for making authorization server better in future.
- **Requesting Party Token**: Token that is related to requesting party is used in authorization process. This token is related to UMA grant. Communication between resource owner and client is done by RPT as a part of request placing. For entities like resource owner, client, requesting party, resource server and authorization server, RPT is unique.
- **Authorization Process**: The main process in the UMA 2.0 protocol among all entities is the role of authorization server that makes decision about RPT issuance to the client. Authorization process is one of the crucial phases in the flow of UMA 2.0 protocol. The issuance of RPT to client decides on the valid conditions on behalf of requesting party. Three important steps that make authorization process simple and smooth are claim collection, authorization assessment process and authorization result determination.

4.4 USE CASES

4.4.1 Healthcare Application

Consider a scenario where Mr. Santosh is a patient acting as a resource owner and is looking for medical consultation from an expert doctor for his abdomen pain. Dr. Sheetalkumar is a special doctor for stomach disorders, and he needs access to Mr. Santosh's all medical records for further diagnosis. Here, access means read scope when an application is concerned. Dr. Sheetalkumar also needs to write scope as he has to update the records with the recent reports (blood test, prescribed tablets and X-ray) of Mr. Santosh. These electronic records are nothing but resources that are updated according to Dr. Sheetalkumar's consent and consultation and are reflected in the application. Dr. Sarika who is the family doctor of Mr. Santosh has already access to these data of Mr. Santosh (Figure 4.3).

Now SMB Healthcare App, as a third-party application, represents Mr. Santosh. Here, he registers to this application and fills all details regarding treatment along with medical reports. SMB Healthcare App also sets some required permissions by using authorization policies, allowing Dr. Sheetalkumar

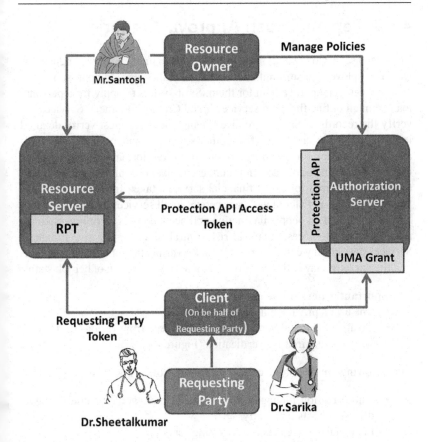

FIGURE 4.3 Use case (healthcare app)

and Dr. Sarika for access control over all the records of Mr. Santosh. Now, Mr. Santosh can grant consent any time to his data just by tapping on the "Share" button or decline the access by tapping on the "Decline" button.

UMA also facilitates to Internet of Things (IoT) considerations. For example, in the above scenario, Mr. Santosh requires that he should be monitored even after his treatment. On his request, Dr. Sheetalkumar suggests a smart medical device to him; it calculates the distance traveled by Mr. Santosh as well as the calories burned by him for further diagnosis. Now this smart device has to be registered with resource server, and also it is protected by authorization server. Eventually, this device will send aggregated data to server, and it will be transferred to the mobile devices or computers through the application so as to monitor Mr. Santosh's parameters for diagnosis. Both Dr. Sheetalkumar and Dr. Sarika will continue to have access to it whenever they log in to their personal devices.

4.4.2 Personal Loan Approval Scenario

Consider a scenario of small scale Industry, due to financial crisis, industry as well as employee get suffered a lot. Eventually, they have to approach financial services that approve loan for them. Sarita wants to apply for a personal loan from an online financial service named Creative Financial Services. To verify the records of Sarita, Creative Finance Services must verify detailed information of Sarita's record from other service providers like Sarita's salary from her employer organization, credit score information and history of it, bank details including account number and balance, and all the information that is required for credit financial services, taken from different hosts where Sarita has registered and has account in all services associated to her. Nowadays, based on people interaction and need, there is an online platform made available to access financial risk central services. Creative Financial Services operator my call to Sarita's bank for verification of her account number, the operator may call to her organization for verification of her job status.

Important aspects to be noted
- The user's privacy must be maintained.
- Creative Financial Services must collect data from various sources, which is required for verification (Figure 4.4).

The following are the main actors in the above flow:

- Authorization manager who works as manager for authorizing different hosts indirectly.
- Financial service acts as requesting service.
- User acts as an authorizing user in flow.
- User bank as a host, for storing all information which are associated with Sarita.
- Job creator of Sarita as a host used for information related to salary.
- Financial risk central services as a host used here for for getting credit information.

4.5 AUTHORIZATION PROCESS

It is considered as one of the important phases in the UMA 2.0 protocol [4] for users to make use of their personal data, and to achieve it in a secure environment for decentralized identity solution, individuals require a reasonable and extensible federated authorization architecture. Access policies can be

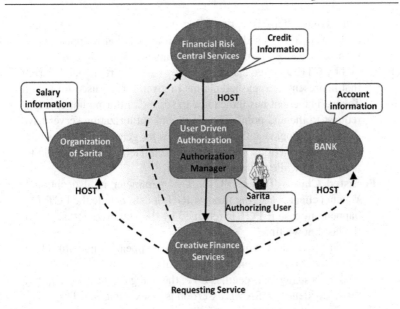

FIGURE 4.4   Use case (personal loan approval flow)

managed by users easily under this type of federation. Users can also grant and take back the access to their data stored on data repositories managed by data management servers. It is the base for extensible federated authorization, which is provided by the protocols and architecture of UMA 2.0. Mainly three activities are involved in authorization process of UMA 2.0 protocol:

- claim collection
- authorization assessment process
- authorization result determination.

4.5.1 Claim Collection

This collection process involves in claim pushing by clients. A client has to request authorization server for RPT for making request to token end points. The client has to send certain parameters, in which some are mandatory, while the others are optional:

Ticket: Request permission ticket which the client has received; this is mandatory in authorization process.

Grant Type: This is also mandatory in authorization process. It has a standard value.

"urn:ietf:params:oauth:grant-type:uma-ticket"

Claims Token: This is optional like making request to token end point. When it is used, then it is mandatory that it appear along with "claim token format" (CTF) parameter. When it is not specified by CTF, it is mandatory to follow base 64 url encoded. String that is present with pushed claims in given format used for Claim Token. The client has the choice to set such information on every request to the end point. The client and authorization server have to mutually decide authorizing policies for general audiences having some restrictions for the claim token. Here security should be maintained while pushing claims in it.

Persisted Claims Token (PCT): This parameter is also optional. When a client is seeking for new RPT, it can include the PCT; this happens in case if PCT is returned by the authorization server in the last known time.

CTF: This parameter is optional. This must appear along with "claim token parameter" when it is used. Like claim token, this format also has a string that specifies CTF; the string used can be a unified resource identifier. SAML assertion is an example of CTFs.

RPT: This optional parameter facilitates authorization server to update the RPT without issuing a new RPT.

Scope: This optional parameter is used to represent the string of spaces with separate values that represent the request scopes.

Authorization Assessment Process: This assessment involves assembling of authorization server and assessing some parameters through which access authorization risk can be mitigated. In this, authorization server assesses a client and checks whether the client is authorized for receiving the request RPT, and based on this, it generates the result. For determining authorization result, authorization server has a hierarchical method of authorization assessment statistics. These statistics are achieved through assessment calculations. Assembling of authorization server is a hierarchical step-by-step process. When first time client was preregistered at authorization server, then first registered scopes are assembled under the scopes that are created.

Like register scope, requested scopes and ticket scopes are also assembled.

Final sets that of requested scopes are determined inside the permission ticket for each resource. Next, assessment and authorization are monitored for every requested scope, where claims policy conditions and other information were input to it. Lastly when each and every scope present in resource satisfies the assessment, it will be added in a set of candidate-granted scopes. Authorization server has certain policy conditions where information gathering in authorization process becomes outdated. As far as authorization process is concerned for future reference.

4.5.2 Authorization Result Determination

The third step in authorization process after authorization evaluation is authorization result determination. In the authorization assessment depending on the scope that is present in resource, it satisfies assessment and authorization server will return success code; if not, it will return error code, an optional PCT or an RPT. There are two types of error codes: "need info" and "request submitted". When this error code is generated, the authorization server gives chance to the client to continue with the same authorization process through permission ticket.

After successful assessment process, the authorization server could modify the parameters in its responses. Both upgrade and PCT are optional parameters in it. Lastly in this assessment, the client is assessed whether it is authorized to receive RPT or NOT, and the result is determined. This task is carried out as authorization server when it receives request regarding RPT, which is sent to the client previously.

UMA 2.0 grant is basically an extension of OAuth 2.0 grant. OAuth capabilities are improved by UMA grant in authorization process. Party-to-party authorization is facilitated by UMA grant. The token associated with UMA grant is requested party token which is OAuth access token [5]. To improve digital access over network, OAuth 2.0 along with grant negotiations and authorization protocol (GNAP) is used. GNAP is associated in IETF. With the help of GNAP, pros and cons of OAuth 2.0 are highlighted. Here, OAuth 2.0 functionalities are compared and represent that how security and privacy can be achieved by well designing authorization servers as token issuers.

4.6 SECURITY ANALYSIS

UMA 2.0 specifications typically rely on the OAuth 2.0 protocol when transport-level security is concerned. Many security considerations can be put together, and countermeasures can be applied to the existing specification draft. UMA 2.0 specifications rely on RFC 6749 of OAuth 2.0 in section 10 along with the entire specification of RFC 6819. So, the developers who use these protocols should first read the recent RFC released for security considerations and should apply countermeasures that are given in the draft of specifications. Security standard requirement for IoT devices is very high compared with others. In Ref. [6], the authors proposed integration of message query telemetry transport (MQTT) along with UMA 2.0. This is used for enhancing the security in an IoT-based application framework [7]. This gives many challenges, which are addressed to solve some real-time authentication scenarios. Many stakeholders are emerging continuously, having different cloud federations as well as different microservices. The access control settlement of such stakeholders under delegation management is considered to make access control more powerful in complex scenarios. This work has proposed DGAC solution models that is dynamic access control solution to overcome security challenges.

Now let's consider some types of security concerns with some scenarios.

4.6.1 PCT and RPT Vulnerability

PCT and RPT roles appear when requesting and accessing mechanism between the entities like client and authorization server at the claims interaction end points where end user requesting party gets registered by the client. There may be the case that client doesn't provide the sequence of queuing references to the authorization server, therefore authorization server doesn't know to handle at the end point which user is present, rather some users may appear directly through the permission ticket. It has some flaws as any end users can be switched remotely by the malicious clients. For example, such client can activate new malicious end user Mr. Suhas by pretending authentic end user named Mr. Sopan. Newly switched malicious end user Suhas may indirectly take permissions and authorize issuance of PCT as Mr. Sopan has already completed authorization process.

Now to overcome such types of threat attacks, some robust strategies need to be introduced, by taking authorization server and response owner in consideration:

- Here from end user requesting party, claims are gathered interactively and use these claims for representation purpose through end user requesting party for exhibiting extreme authentication is executed in process. This may help to legitimate a threat.
- To fulfill the authorization server's policy conditions, it is required that claim must be made freshly every time. Only if the claims are fresh, it can easily satisfy policy conditions of authorization server.
- One important consideration regarding RPT is that access token must be shared with authorization server and resource server only. This will be shared by a legalized client. And it is nowhere

shared to any of the clients or any of the requesting party. As a requesting party, who is malicious, who can steal RPT, that the resource owner doesn't.

- To keep the PCT confidential while in transportation as well as in storage, there is a possibility of client and authorization server to not to share PCT with any one and each other. The PCT client bonding with each other will be maintained by authorization server.

4.6.2 Cross-Site Request Forgery Attack (CSRF)

Figure 4.5 illustrates cross-site forgery attack, in which the attacker obtains the details of a victim and sends a request to the victim pretending that the request is sent from a bank. In this flow, at first, victims log in to their bank account by filling their account credentials (username and password). After that, the bank sends a validation token if the credentials sent by the victims matched. Now the hacker sends a forged request pretending as a request sent by the bank as a legitimate communication. This request will be forwarded by the victim unknowingly to their bank. As that request was forged, it will be executed by the bank using a previously assigned token. Eventually, the hacker takes charge of the victim account and transfer all funds to his account. Here all this happened due

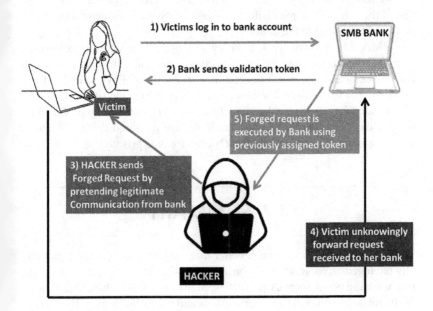

FIGURE 4.5 Cross-site request forgery attack (without UMA 2.0 protocol)

to access violence initiated by the attacker, and the method used here for claim gathering is in an interactive way from end user requesting party. This leads to potential cross-site request forgery (CSRF) attacks.

Now, to avoid this, the UMA 2.0 protocol has introduced "state" parameter value comparison. Claim interaction end point of authentication server is the target of the attacker in CSRF attack, where there are chances of gaining the authorization for access by using malicious client, by keeping end user requesting party or client as the same [8]. Till date, a number of countermeasure methods are proposed, considering current methods of best practices are also not secured. Here the authors targeted OAuth Cross-Site Request Forgery (OCSRF) scenario as a missed attack. After studying numerous OCSRF attacks, OAuth has proposed a strategy. This is dependent on the browser of victim.

It is important that authorization server must have a system to protect claims interaction end point, ensuring that without requesting party, malicious client could not gain the authorization. "State" parameter is used to overcome such scene, while in redirection requesting party to the claim interaction end point. "State" parameter value can't be traceable to the hacker. Binding value is associated with "state" parameter, and is sent by authorization server when it redirects back the requesting party. Here comparison of value of "state" parameter is carried out. When this "state" parameter was sent to initial redirection request with recently received "state" parameter, in case if the values don't match then such invalid "state" parameter will results to an error message to the attacker.

UMA 2.0 protocols seek a strong security mechanism and specifications to countermeasure attacks, as security concerned for data privacy in framework of authorization life cycle, in Ref. [9], this policy profile has been introduced in this. This is for data privacy in authorization life cycle. Using Abbreviated Language for Authorization (ALFA) in the policy profile, data privacy can be achieved. Here the authors have proposed three techniques with extension of prior existing models used for data privacy.

4.7 SUMMARY

To summarize UMA, it is much better that earlier versions when conceptually put together. More security concerns are taken into considerations, eventually adding better security properties. Many contributors have added more drafts related to token, string, permissions and access delegation. By looking at the new specifications, it seems to be robust protocol on top of OAuth

2.0. And it is right indication that this specification is moving toward right direction. Implementers have started implementing it and obtaining the best feedback from clients, as redirection and other security parameters are concerned. UMA 2.0 protocol can easily be implemented on the top of OAuth 2.0., having extended specifications with UMA grant added. While delegation of access, it has a strong mechanism to authorize client or requesting parties, by authorization assessment process.

REFERENCES

1. Siriwardena, P. (2014). User Managed Access (UMA). In Advanced API Security (pp. 155–170). Apress, Berkeley, CA.
2. Siriwardena, P. (2020). UMA Evolution. In Advanced API Security (pp. 377–396). Apress, Berkeley, CA.
3. E. Bertin, D. Hussein, C. Sengul and V. Frey. (2019). Access control in the Internet of Things: a survey of existing approaches and open research questions. *Annals of Telecommunications*, Springer 74(7), 375–388.
4. T. Hardjono. (2019). Federated authorization over access to personal data for decentralized identity management. *IEEE Communications Standards Magazine*, 3(4), 32–38.
5. Imbault, F., Richer, J., & Parecki, A. (2021). Managing authorization grants beyond OAuth 2. Open Identity Summit 2021.
6. K. S. Aloufi and, O. H. Alhazmi. (2020). A hybrid IoT security model of MQTT and UMA. *Communications and Network*, 12(04), 155.
7. D. Preuveneers and Joosen, W. (2019, June). Towards multi-party policy-based access control in federations of cloud and edge microservices. In *2019 IEEE European Symposium on Security and Privacy Workshops (EuroS&PW)* (pp. 29–38). IEEE, Stockholm, Sweden.
8. M. Benolli, S. A. Mirheidari, E. Arshad and B. Crispo. (2021, July). The full gamut of an attack: An empirical analysis of OAuth CSRF in the wild. In *International Conference on Detection of Intrusions and Malware, and Vulnerability Assessment*, pp. 21–41, Springer, Cham.
9. S. Bandopadhyay, T. Dimitrakos, Y. Diaz, A. Hariri, T. Dilshener, A. La Marra and A. Rosetti. (2021, September). DataPAL: Data protection and authorization lifecycle framework. In *2021 6th South-East Europe Design Automation, Computer Engineering, Computer Networks and Social Media Conference (SEEDA-CECNSM)*, pp. 1–8, IEEE.

Conclusions

5

Due to advancements in technologies, the use of online services has increased. More than 380 websites are created every minute and are placed on different domains; therefore, the role of security in every aspect of the network becomes more challenging these days. Web applications require robust frameworks to provide more security. Vulnerability becomes a key security challenge these days. In the context of a large number of web applications and to sign up those with required details requires more time. Maintaining a user's account credential is also one of the challenges to users as well as cloud platform services dealing with web applications.

In view of this, Chapter 1 describes how Internet of Things (IoT) has evolved from the Internet and became the future of the Internet, and the challenges faced while connecting the things with the Internet. This chapter describes information and telecommunication technology and its standardization, and how Information and Communication Technology (ICT) and its standardization help IoT to grow more in both technological and economical perspectives. Also, bringing many different organizations to one platform and setting the standard for ICT will help to develop both developing and developed countries. This chapter presents the convergence of IoT with different technologies, and there is a discussion about some examples of cool applications and technology-based convergence of IoT. The next part of this chapter presents industrial revolution four or Industry 4.0, and a detailed description of all the four revolutions or industry shifts and the Industry 4.0 standard and reference model is provided. There is some discussion about how this standard will help us to grow technologically and economically or how Industry 4.0 technology will grow the economy of the world. This chapter also presents a discussion on some key challenges and issues faced by IoT developers or IoT systems while designing and developing IoT devices or systems or platforms or frameworks.

In Chapter 2, an overview of authentication, authorization and access control mechanisms is presented. As many organizations are keeping their data on the cloud and providing access to these data through web services, users can access these data from any place at any time through any device.

Most of the communication is done through wireless mode, and hence, it is prone to security attacks. Before communication, both end parties should know whether they are communicating to the legitimate partner. Hence, authentication plays an important role in securing the services. Furthermore, authorization and access control are important to allow/deny access to the legitimate user based on the policies. The difference between authentication, authorization and access control is also discussed in this chapter. The last part of this chapter discusses the various mechanisms of access control in detail. In this digital era, there is a need for a decentralized, privacy-aware access control mechanism. In view of this, the various techniques like Decentralized Identifier (DID), Self-sovereign identity (SSI), Verifiable Credentials (VC), are also discussed in this chapter.

OAuth2.0 protocol and an extended version of it as a UMA 2.0 provide faster access control through the smarter token exchanging method. The main focus of the book is that how we can use OAuth 2.0 and UMA 2.0 framework for redirecting/connecting to third-party applications with minimal user credentials that are provided by authorization servers or many times resource servers. Eventually, users can maintain their number of accounts on various web applications, portals or any social sites only by maintaining a single primary account. Chapters 3 and 4 present a detailed discussion of OAuth 2.0 and UMA 2.0, respectively. An overview of OAuth2.0 and UMA2.0 is presented with the help of use cases and a detailed security analysis.

Index

A

Abbreviated Language for Authorization (ALFA) 68
abstract protocol flow 38
access token 35
application programming interface (API) 33
Association of Radio Industries and Businesses (ARIB) 3
Attribute-based access control model (ABAC) 26
authentication 24
authorization 24
authorization assessment process 64
authorization code grant 42
authorization framework 35
authorization result determination 65
authorization server 35

C

claim collection 63
claim token 60
claim token format (CTF) 64
claim token parameter (CTP) 64
client credentials grant 42
content security policies (CSP) 50
cross-site request forgery attack (CSRF) 67

D

data integrity attack 22
decentralized identities (DID) 29
denial of service attack 23
DNS spoofing 50

E

European Telecommunications Standardization Institute (ETSI) 3

F

Fast Identity Online (FIDO) Alliance 25
federated authorization 54
fragment of URL 39

G

global ICT Standardization Forum for India (GISFI) 3
grant negotiations and authorization protocol (GNAP) 65

H

HTTPS 33

I

IETF OAuth Working Group 35
implicit grant 42
Information and Communication Technology (ICT) 3
International Telecommunication Union (ITU-R) 3
internet of things (IoT) network 34
internet of things 1
interoperable access token (IAT) 42

J

JSON Web Token (JWT) 33

M

machine-to-machine (M2M) 2
Man-in-the-Middle (MitM) attack 23
message query telemetry transport (MQTT) 66
MyDataChain framework 46

N

non-interactive zero-knowledge (NIZK) 46

O

OAuth Cross-Site Request Forgery (OCSRF) 68
OpenAM 25
Open Authorization (OAuth 2.0) 33
OpenID 25
Over The Top (OTT) 1

P

permission ticket 54
Persisted Claim Token (PCT) 60
phishing 21
phishing attacks 49
Policy Administration Point (PAP) 25
Policy Decision Point (PDP) 25
Policy Enforcement Point (PEP) 25
Policy Information Point (PIP) 25
protection API access token (PAT) 58

R

ransomware 21
Reference Architecture Model for Industry
4.0 (RAMI 4.0) 7
requesting party 55
requesting party token (RPT) 57
resource owner 35
resource owner password credential grant 42
resource server 35
RFC 6749 33
RFID 7
role-based access control model (RBAC) 26

S

self-sovereign identity (SSI) 29
Sybil attack 21

T

Telecommunications Industry
Association (TIA) 3
third-party application 37
token binding mechanism 36
traditional client server authentication
model 36

U

UMA-federated authorization
recommendation 57
UMA grant 56
uniform resource identifier (URI) 43
uniform resource locator (URL) 39
user driven authorization 63
user-managed access (UMA) 53

V

verifiable credentials (VC) 29
VOAuth 50

W

Wireless Local Area Network (WLAN) 4

Printed in the United States
by Baker & Taylor Publisher Services